Charles Jewett, Harold Flagg Jewett

Outlines of Obstetrics

A Syllabus of Lectures Delivered at the Long Island College Hospital

Charles Jewett, Harold Flagg Jewett

Outlines of Obstetrics
A Syllabus of Lectures Delivered at the Long Island College Hospital

ISBN/EAN: 9783337163556

Printed in Europe, USA, Canada, Australia, Japan

Cover: Foto ©berggeist007 / pixelio.de

More available books at **www.hansebooks.com**

OUTLINES

OF

OBSTETRICS

A SYLLABUS OF LECTURES
DELIVERED AT THE LONG ISLAND COLLEGE HOSPITAL

BY

CHARLES JEWETT, A.M., M.D.,

PROFESSOR OF OBSTETRICS AND PEDIATRICS IN THE COLLEGE, AND **OBSTETRICIAN**
TO THE HOSPITAL.

EDITED BY

HAROLD F. JEWETT, M.D.

———————————

PHILADELPHIA:

W. B. SAUNDERS,
925 WALNUT STREET.

1894.

NOTE.

THIS syllabus is intended as an aid to the study of Obstetrics in both the didactic and the practical work of the college course. Its main aim is to help the student in securing a classified knowledge of the outlines of his subject, which, it is believed, should be the first step in the pursuit of any branch of learning. This once accomplished, his progress will no longer be difficult. Upon a well-ordered framework of general facts and principles further acquisitions classify themselves, and a complete and systematic knowledge of the subject becomes a matter of comparatively easy growth.

The hope is entertained that the work may be found of some value to the practitioner also as a convenient handbook for reference.

330 CLINTON AVE., BROOKLYN, N. Y.,
November, 1893.

CONTENTS.

OUTLINES OF OBSTETRICS.

THE FEMALE GENITAL ORGANS.

The Pudendum. The term pudendum is applied to the external sexual organs of the female collectively. It includes the mons Veneris, the labia majora and minora, and the clitoris. Vulva is used in a similar sense, but, strictly speaking, includes only the labia.

The Mons Veneris is a fatty cushion overlying the pubic bones. It is continuous with the hypogastrium above, below it merges into the labia majora; laterally its boundaries are the groins. In addition to fat it contains fibrous and elastic tissue. Its surface abounds in sebaceous and sweat glands, and at puberty becomes invested with a growth of hair.

The Labia Majora are two prominent, rounded, fleshy folds springing from the mons Veneris, and extending downward and backward to a point about one and a quarter inch in front of the anal orifice. They are symmetrically placed on either side of the median line, and in full development they lie in contact with each other in the young nullipara. When shrunken from loss of fatty tissue, the labia minora protrude between them. They are thickest in front and taper from before backward. The point of contact in front is the anterior and that behind the posterior commissure.

2

Their covering is skin. The outer surfaces are supplied with hair; the inner resemble mucous membrane, but are sparsely covered with fine hairs. Both surfaces abound in sebaceous glands. Their internal structure is made up chiefly of elastic and adipose tissue, and includes a rich venous plexus. The remains of the canal of Nuck may sometimes be traced into them on either side. They are the analogue of the scrotum in the male.

The **Labia Minora** or **Nymphæ** are two thin folds of muco-cutaneous tissue lying obliquely upon the inner surfaces of the labia majora. They are widest toward their anterior extremities, narrowing gradually from before backward. Anteriorly each subdivides into two subsidiary folds, the superior folds uniting in front of the clitoris to form the prepuce, the inferior folds forming, by their junction below the glans, the frenum of the clitoris.

The **Fourchette** or **Frenulum Vulvæ** is a transverse fold uniting the labia minora posteriorly.

The **Fossa Navicularis** is the boat-shaped surface which appears between the fourchette and the hymen when the labia are separated.

The **Rima Pudendi** is the median cleft between the labia of the right and left sides.

The **Clitoris** is the analogue of the penis. It lies in the median line just below the anterior vulvar commissure, concealed behind the mucous membrane. It has two corpora cavernosa and a glans, but no corpus spongiosum, and is imperforate. Continuous with the corpora cavernosa of the clitoris are the crura by which it is attached to the ischio-pubic rami. The length of the clitoris during erection is about one inch.

The Vestibule is the triangular space bounded laterally by the labia minora and below by the margin of the vaginal orifice. Its covering is mucous membrane. At its apex is the glans clitoridis. At the center of its base, or immediately above it, is the meatus urethræ in the center of a small tubercle or prominence. The meatus lies three-fourths of an inch below the glans clitoridis, and one inch above the fourchette in the nullipara.

The muco-cutaneous line runs along the lateral borders of the vestibule and passes backward around the vaginal orifice just without the base of the hymen.

The arterial supply of the pudendum is chiefly from the internal pudic.

The veins. The labia, clitoris, and urethra abound in erectile tissue rich in venous plexuses. The *bulbi vestibuli* are two leech-shaped masses of veins about an inch in length, and situated one on either side of the mesial line behind the labia, opposite the vaginal orifice and the base of the vestibule. They lie between the bulbo-cavernosus muscle and the vaginal wall, immediately in front of the triangular ligament. They communicate freely with the veins of the labia, the vagina, the perineum, the glans clitoridis, and with other neighboring venous plexuses.

The lymphatics of the labia majora, minora, and clitoris terminate in the superficial inguinal glands.

The nerve supply of the pudendum is derived chiefly from the internal pudic. It is especially abundant in the clitoris and the labia minora.

The pudendal glands. Sebaceous glands abound upon the skin surfaces, and especially on the nymphæ. About half a dozen muciparous glands are to be found in the vestibule, grouped about the meatus urethræ.

The vulvo-vaginal glands, glands of Bartholin or *Duverney,* are two reddish-yellow bodies varying in size from a pea to an almond, lying one on each side of the posterior portion of the vagina, behind the anterior layers of the triangular ligament. They lie partly behind the lower extremities of the bulbi vestibuli. Their ducts, about a half-inch in length, run along the inner aspects of the bulbi vestibuli, opening just without the base of the hymen at the sides of the vaginal orifice. Their secretion is poured out freely under sexual excitement and during labor.

The **Hymen** appears as a septum partially occluding the vaginal orifice when the labia are drawn apart. When at rest, it protrudes as a loose fold in the vulvar fissure. According to Budin, it is a thinned-out prolongation of the vagina itself. Its most common form is that of a crescent, situated at the posterior margin of the vaginal orifice, with its concavity looking forward. It may, however, be annular, or may occupy the entire vaginal orifice, being either imperforate or cribriform— perforated with holes—or may have a single central opening with a fimbriated edge. It is usually torn at the first sexual approaches. An untorn hymen is not, however, an infallible mark of virginity, nor is a torn hymen necessarily evidence that sexual intercourse has been practised.

The **Carunculæ Myrtiformes** are the remains of the hymen torn in labor by the passage of the child. They appear as minute fleshy tubercles skirting the vaginal orifice or its posterior margin.

The **Vagina** is that part of the genital tract between the uterus and the pudendum. It is a collapsed tube, its anterior and posterior walls lying in contact.

Relations. Above, it is attached to the uterine cervix at about the middle of its length, the lower portion of the cervix projecting into the vagina nearly at a right angle; below, it is attached to the ischio-pubic rami. Its posterior wall, at its upper extremity, is in relation with the peritoneum, at its lower with the perineal body; at its middle portion it is loosely connected with the rectum. The upper half of the anterior wall is loosely attached to the bladder; the lower half is intimately connected with the urethra.

That part of the canal immediately about the vaginal portion of the cervix uteri is termed the fornix or roof of the vagina.

The length of the anterior wall is two and a half inches; of the posterior, three inches or a little more. The walls are, however, extremely distensible, and they become permanently relaxed in parous women.

Shape. The vagina, when distended, has a conoidal shape, the orifice corresponding to the smaller end of the cone.

Structure. Three coats. 1. External, a loose layer of connective tissue. 2. Middle, muscular coat, comprising circular and longitudinal bundles of unstriped muscular fiber. 3. Internal, mucous membrane.

A bundle of voluntary muscular fibers—the *sphincter vaginæ* (the bulbo-cavernosus muscle)—encircles the vaginal orifice.

The columnæ vaginæ are two median ridges, one on the anterior and one on the posterior wall in the median line, at the lower portion of the vagina. The *cristæ* are transverse ridges running outward from the columnæ vaginæ. They are most marked near the orifice and upon the anterior wall. The cristæ are more or less effaced by childbearing or by catarrhal inflammation of the vagina.

The vaginal epithelium is of the squamous variety of epithelium.

The arterial supply of the vagina is chiefly from the vaginal, but also from branches of the uterine and of the pudendal arteries. Its arteries anastomose with the vesical and rectal arteries.

The veins correspond, but they first form a plexus around the vagina. They communicate freely with the hemorrhoidal, vesical, pudendal, and pampiniform plexuses.

The lymphatics of the lower fourth of the vagina join with those of the pudendum, terminating in the inguinal glands. Those from the remaining portion of the vagina unite with the vesical and cervical lymphatics and empty into the hypogastric glands.

The nerves are from the fourth sacral and the pudic of the spinal system, and from the hypogastric plexus of the sympathetic.

The mucous glands are chiefly confined to the lower portion of the vagina.

The Urethra.—Intimately connected with the lower portion of the anterior vaginal wall is the urethra. Though not a genital organ, it is of obstetric interest, and is therefore described.

Situation. The urethra passes backward beneath the pubic arch to the bladder, nearly in a straight line parallel to the plane of the pelvic brim. Its lower three-fourths is inseparable from the anterior vaginal wall.

Size. The length of the urethra is one and a half inch; its average diameter, one-quarter inch. It is largest at the vesical end, smallest at the meatus, and is very distensible.

Structure. It has two muscular coats and a mucous membrane.

Skene's glands are two tubular glands about three-fourths of an inch in length, to be found in the muscular wall of the urethra at its lower end, near its floor, one on either side of the median line. Their orifices lie just within the meatus urethræ.

The Uterus.

Situation, in the cavity of the pelvis, between the bladder and the rectum, a little nearer the sacrum than the pubes. Its upper border is nearly in the plane of the pelvic brim, its lower border about opposite the tip of the sacrum. The average direction of its long axis is nearly perpendicular to the plane of the pelvic brim. Its position, however, is variable.

Shape, pyriform—larger end upward—flattened from before backward, its posterior and upper borders convex, its anterior surface nearly flat.

Size, a. Nulliparous uterus, average measurements, nearly one inch thick, one and a half inch wide at the fundus and two and a half inches long. *b.* Parous uterus, approximately one inch thick, two inches wide, and three inches long.

Weight, nulliparous, about one ounce; parous, one and a half ounce.

Regional divisions. The uterus presents three divisions —the cervix, the body, the fundus.

The cervix is approximately the lower half of the uterus in the nulliparous, the lower third in the parous woman.

The fundus is the part above the Fallopian tubes.

The body is the part between the cervix and the Fallopian tubes.

The isthmus is the slight constriction at the junction of the cervix and the body.

Divisions of the cervix.

a. The infra-vaginal portion, or portio vaginalis, is that part of the cervix below the vaginal roof.

b. The supra-vaginal portion is that part between the portio vaginalis and the isthmus.

Uterine cavity.

a. The cavity of the body is somewhat triangular in shape in the nullipara, its anterior and posterior walls lying in contact. It has three openings, one communicating with the cervical canal and one with each of the Fallopian tubes.

b. The cavity of the cervix is somewhat flattened from before backward, and is laterally elliptical, thus having an irregular fusiform shape.

The os internum is the upper orifice of the cervical canal, and is about one-tenth inch in diameter;

The os externum is the lower orifice, a little larger than the os internum.

Structure.

I. *The mucous membrane. a.* Body. About one-twenty-fifth of an inch thick; its epithelium is of the ciliated columnar variety, the cilia propelling toward the fundus. It abounds in tubular glands, many of which are bifurcated. These are obliquely placed and lined with ciliated epithelium. They secrete an alkaline mucus. There is no submucous layer. *b*. Cervix. Thinner, of a paler color and firmer than that of the body.

Arbor vitæ is a term applied to the median longitudinal ridge, on the mucosa of each wall of the cervix, together with the transverse folds or rugæ running obliquely outward and upward from it.

Nabothian glands. In and between the rugæ of the arbor vitæ are mucous follicular glands known as the Nabothian glands. The secretion of the cervical glands is a tenacious mucus having an alkaline reaction.

The epithelium of the cervical canal is of the ciliated columnar variety almost to the os externum in the adult. The epithelium of the external surface of the portio vaginalis is squamous like that of the vagina.

II. *The musculature* of the body constitutes the greater part of the thickness of the uterine wall. Its fiber is of the unstriped variety.

Layers.

a. Outer, very thin, continuous with the muscular layers of the Fallopian tubes, the ovarian, round, broad, utero-sacral and utero-vesical ligaments.

b. Middle, comprising the bulk of the uterine muscle, a meshwork of longitudinal and transverse bundles.

c. Inner, circular layer, extremely thin, surrounding the orifices of the Fallopian tubes and the os internum.

The cervix consists mainly of connective tissue.

III. *The peritoneal coat.* The peritoneum covers about two-thirds the length of the uterus in front and posteriorly extends down beyond the uterus over about one inch of the vagina.

Nulliparous and parous uterus. In the *nulliparous uterus* the corporeal cavity is triangular, fundus nearly flat, cervix conical, and the os externum a mere dimple. In the *parous uterus* the cavity is oval, the fundus dome-shaped, the cervix cylindrical, and the os externum a transverse slit, with the lips more or less fissured.

Ligaments of the uterus.

a. The broad ligaments. The pelvic peritoneum dips down posteriorly into the lesser pelvis, is reflected over one inch or more of the upper part of the posterior vaginal wall, covers the posterior surface of the uterus, passes over the fundus and invests the anterior uterine surface to the isthmus; thence it is again reflected upward and over the bladder. The uterus thus lies between the layers of a transverse fold of peritoneum. The lateral portions of this transverse fold, stretching from the uterus to the sides of the pelvis, in front of the sacro-iliac joints, form the broad ligaments. The two layers of each broad ligament are practically in apposition except at their junction with the pelvic floor and with the pelvic walls. The ovarian ligament, the Fallopian tube, and the round ligament are enveloped in subsidiary folds of the broad ligament.

The infundibulo-pelvic ligament is that part of the superior border of the broad ligament on each side, extending from the Fallopian tube to the pelvic wall.

b. The utero-sacral folds are two semilunar folds of peritoneum, enclosing unstriped muscular fiber and connective tissue, which pass one on each side of the rectum from the lower portion of the sides of the uterus to the second bone of the sacrum. In the nulliparous woman they spring from the uterus at the level of the os internum ; in the parous, from points somewhat above the os internum. These folds are also known as the folds of Douglas, and the space between them as Douglas's pouch, or cul-de-sac.

c. *The utero-vesical folds* are two folds of peritoneum extending from the uterus to the bladder and forming the lateral borders of the utero-vesical space. They contain a few muscular fibers.

d. *The round ligaments* are two muscular bundles which spring from the angles of the uterus in front of the Fallopian tubes, and pass forward through the inguinal canals to blend with the structures at and immediately below the external ring. They consist of striped and unstriped muscular fibers. Their length is four to five inches.

The arteries of the uterus are the hypogastric and ovarian. They pass to it between the folds of the broad ligament on either side. The hypogastric artery approaches it just above the vaginal junction, the ovarian at the level of the cornua. A branch of the ovarian artery descends along the lateral border of the uterus to communicate with the hypogastric. Another branch supplies the fundus and anastomoses with its fellow of the opposite side. The circular artery surrounds the cervix at the isthmus, uniting the arteries of the opposite sides with each other. The arteries of the uterus are remarkable for their free anastomosis and tortuous course. Arterial tufts are given off to the lateral borders, whose branches form spirals within the uterine

wall.   They end in a meshwork of capillaries about the
utricular glands.

The veins.   The uterine plexus of veins lies immediately beneath the peritoneal coat of the uterus and
extends between the folds of the broad ligament.   It
communicates with large sinuses in the middle muscular
coat which are encircled by muscular bundles.   Its outlet
is the hypogastric vein and the pampiniform plexus.

The lymphatics are very numerous in the body of the
uterus, and communicate with the lymph spaces of the
mucous membrane and the muscular coat.   They form
an intricate network immediately beneath the peritoneal
coat of the uterus and the Fallopian tubes.   The uterine
lymphatics are fully developed only during pregnancy.
The lymphatics of the fundus follow the course of the
Fallopian tubes to the ovary, thence to the lumbar
glands.   Those of the body pass outward in the broad
ligament to the iliac glands.   The cervical lymphatics
unite with those from the upper part of the vagina
and empty into the hypogastric glands alongside the
rectum.

The nerves are derived chiefly from the sympathetic
system, from the inferior hypogastric and spermatic
plexuses.   The uterus also receives filaments from the
first, second, and third sacral nerves.   The uterine nerves
terminate in part in the nuclei of the muscle cells.

The Fallopian Tubes or **Oviducts** are two narrow
tubes, one running outward from each horn of the
uterus and communicating with the uterine cavity.   The
outer portion of each tube takes a tortuous course, partially surrounding the ovary.   The length of the tubes
is from three to five inches.

Divisions.

a. The isthmus is the portion of the tube next the uterus. It is an inch or more in length and one-eighth of an inch in diameter.

b. The ampulla is the dilated portion of the tube next beyond the isthmus, about one-third of an inch in diameter.

c. The fimbriated extremity, pavilion or *infundibulum,* is the free trumpet-shaped end of the tube, the margin of which is fringed with a number of irregular processes called fimbriæ. Here the tube expands to about three-fourths of an inch in diameter.

The fimbria ovarica is the special fimbria, a little larger than the others, which is attached to the ovary.

The ostium uterinum barely admits a bristle—$\frac{1}{25}$ inch in diameter.

The ostium abdominale is of the size of a small goose-quill.

Structure. Three Layers.

1. *Outer* or *peritoneal coat,* which invests two-thirds the circumference of the tube. A subperitoneal layer of connective tissue contains a rich plexus of blood-vessels.

2. *Middle* or *muscular coat,* composed of an inner circular and two outer longitudinal layers of unstriped muscular fiber.

3. *Inner* or *mucous coat,* lined with ciliated columnar epithelium and very vascular. Except in the intra-mural portion of the tube, the mucous membrane is disposed in longitudinal folds, which become extremely complex in the ampulla. The motion of the cilia propels toward the uterus.

The arteries of the Fallopian tube are branches of the ovarian.

The veins open into the pampiniform or ovarian plexus lying between the folds of the broad ligament below the tube.

The lymphatics unite with those from the fundus of the uterus and from the ovary and terminate in the lumbar glands.

The nerves are derived from the inferior hypogastric plexus on each side.

The Ovaries are two in number. They correspond to the testes of the male.

Situation, one on each side of the uterus, an inch or more below the level of the ilio-pectineal line and the same distance from the uterus; yet they have great mobility within normal limits. Each ovary is attached by its anterior straight border to the posterior surface of the anterior fold of the broad ligament, projects through the posterior fold, and is connected with the corresponding horn of the uterus by the ovarian ligament.

Shape. The usual shape of the ovary is a flattened ovoid; its free border is convex; the anterior edge is nearly straight. This straight border is the hilum. The ovary is thinnest at the hilum, thickest at the convex border. The inner end is narrower, pointed, and merges into the ovarian ligament; the outer is more obtuse and bulbous. The shape, however, is variable.

Size, about one and a half inches in length by three-fourths in width and one-half in thickness, but is variable. The average normal weight in the nullipara is eighty-five grains. The size increases during menstruation, also in early pregnancy.

Structure.

A. External. In early life the external surface is smooth, like an almond. Later in life, after puberty, it becomes uneven, acquiring a wrinkled appearance, owing to the cicatrices from rupture of Graafian follicles. In very old age it again becomes smooth. Its epithelium is columnar and non-ciliated—the germinal epithelium of Waldeyer.

B. Internal.

a. The *stroma* contains some unstriped muscular fiber in addition to connective tissue.

b. The *tunica albuginea* is a dense layer of stroma immediately underlying the germinal epithelium of the ovarian surface.

c. The *zona parenchymatosa* is the cortical portion of the ovary; it has a grayish color.

d. The *medullary zone,* or *zona vasculosa,* is the portion about the hilum; it is reddish in color. Here enter the bloodvessels, nerves, and lymphatics.

The ovarian ligament is a muscular band one-fifth inch in width, which extends from the inner end of the ovary to the horn of the uterus, joining it immediately behind and below the attachment of the Fallopian tube. Its length is about an inch.

The arterial supply of the ovary is derived from branches of the ovarian artery which enter at the hilum.

The veins emerge from the hilum and empty into the pampiniform plexus.

The lymphatics, with those of the tube and fundus uteri, empty into the lumbar glands.

The nerves are derived from the inferior hypogastric plexus.

The Graafian Follicles are the sacs in which the ova are developed. The follicles are developed from the germ epithelium of the ovarian surface, which becomes imbedded in the stroma by the outgrowth of connective tissue. They are most numerous in the cortical layer. Each follicle contains generally but one ovum. The number of rudimentary Graafian follicles at birth is 40,000 or more in each ovary. At any time during the childbearing period ten or twenty Graafian follicles may be found in different stages of development upon the ovarian surface. The size of a mature Graafian follicle is $\frac{1}{100}$ to $\frac{1}{16}$ inch in diameter.

Structure of a Graafian follicle. The constituent parts of a Graafian follicle are: 1. The tunica fibrosa. 2. The tunica propria. 3. The tunica (membrana) granulosa, a multiple layer of cylindrical epithelium. The discus proligerus or germinal eminence is a heaped-up mass of cells of the membrana granulosa at one side, containing the ovule. 4. The liquor folliculi, a clear albuminous fluid, para-albumin.

MALFORMATIONS OF THE UTERUS.

Uterus unicornis. One lateral half absent; has generally but one Fallopian tube.

Uterus didelphys. A bifid uterus; each lateral half forms a distinct organ.

Uterus bicornis. The lateral halves are distinct above, united below—the upper part of the uterus is bifid.

Uterus cordiformis. The fundus presents an antero-posterior median sulcus.

Uterus septus. The uterine cavity is divided, wholly or partially, into two lateral cavities by a median partition.

PHYSIOLOGY OF PREGNANCY.

OVULATION. MENSTRUATION. CONCEPTION.

Ovulation is the process by which the ovum is matured and discharged from the ovary. It occurs, as a rule, once in twenty-eight days, during the period of functional activity of the ovary; is attended generally in the human subject with a bloody discharge from the uterus—menstruation. Yet ovulation may occur without menstruation, and menstruation without ovulation. Usually but a single follicle ruptures at each epoch, sometimes two or more.

Menstruation is a periodic congestion of the female genital organs, attended with a bloody discharge from the uterus—*the menses.* The endometrium undergoes partial or possibly complete exfoliation and subsequent renewal.

The constituents of the menstrual flow are blood, shreds of endometrium, uterine and vaginal secretions. The amount is four to six ounces; the average duration of the flow, three or four days; the interval between the menstrual epochs is usually twenty-eight days.

Puberty is the period of sexual maturity, and is marked in the female by the first onset of ovulation and menstruation.

The age of puberty is usually about the fifteenth year. It varies with race, climate, and other influences, occurring in exceptional cases as early as the tenth or as late as the twentieth year of age.

The menopause, or the final cessation of menstruation and of the capacity for childbearing, occurs in most women at about the age of forty-six years. Occasional variations of ten years or more on either side of this limit are possible.

Phenomena attending the Rupture of a Graafian Follicle.—An increase of fluid contents takes place from increased vascularity. Loops of bloodvessels are projected into the cavity of the follicle. Contiguous portions of the ovary, and to a certain extent its whole structure, exhibit a similar increase in vascularity. The follicle now appears as a bright-red spot on the surface of the ovary.

Absorption of overlying ovarian structure takes place owing to increasing pressure of the liquor folliculi. The follicle finally ruptures and discharges its contents, an effusion of blood taking place into the follicle after rupture. Apparently the ovule is floated into the pavilion of the tube by a stream of serum propelled by the cilia of the fimbria ovarica. It is propelled through the Fallopian tube partly by ciliary motion, and in the narrower portion of the tube partly by muscular action. Rarely it happens that the ovum finds its way across the pelvic cavity and into the opposite Fallopian tube.

The Ovum is primarily a nucleated cell developed from the germ epithelium covering the surface of the ovary. The diameter of the ovum at maturity is $\frac{1}{120}$ inch.

The parts of the mature ovum are: 1. *The vitelline membrane.* 2. *The vitellus,* or *yolk,* a mass of oleo-albuminous matter containing shining granules. 3. *The germinal vesicle,* the nucleus of the cell, $\frac{1}{700}$ inch in diameter, situated toward one side of the yolk near its surface. 4. *The germinal spot*, the nucleolus of the cell, a dark, granular spot, about $\frac{1}{3000}$ inch in diameter, within the vesicle.

The female pronucleus. Out of the germinal vesicle, or that part of it which remains after certain developmental changes, is formed the female pronucleus. As will be seen later, the fusion of the female with the male pronucleus is the essential fact in the process of fecundation.

The Corpus Luteum is the body formed in the ovary by the retrograde metamorphosis of the Graafian follicle after rupture.

The corpus luteum of menstruation attains its full development in two to four weeks, and becomes reduced to a mere cicatrix by the end of about two months.

The corpus luteum of pregnancy grows six or seven weeks, then remains stationary to the end of the fourth month; subsequently it retrogrades slowly until delivery and becomes an insignificant cicatrix by the end of a month after childbirth.

CONCEPTION: IMPREGNATION.

Impregnation, or **Conception**, is the fructification of the ovum by union with the fecundating element of the male—the spermatozoön. Insemination is the act of depositing the seminal fluid in the female genital tract.

The seminal fluid is a glutinous, alkaline, albuminous fluid, heavier than water, the combined product of the testicles, the prostate, and Cowper's glands. From one to three drachms is ejaculated during the orgasm. Its chemical constituents are water, proteids, phosphates, and fats; its microscopical elements, spermatozoa and phosphatic crystals.

The spermatozoa are microscopic bodies resembling tadpoles in shape. Each consists of a flattened ovoid head (cell nucleus) and a long thread-like tail. The filiform tails maintain a constant lashing motion, due to amœboid movements of protoplasm, as long as the spermatozoa retain their fecundating power. The length of a spermatozoön is $\frac{1}{600}$ to $\frac{1}{400}$ inch.

Vitality of spermatozoa. Under favorable conditions, within the genital passages of the female, the spermatozoa as well as the ovum retain their vitality for an unknown period of time. In the human species they have been known to live for eight days. They may undoubtedly retain their fecundating power for a much longer time.

They are destructible by extreme heat or cold. The seminal elements of man retain their power of motion, however, between the temperatures of 5° and 116° F. They are destroyed by acids, by numerous other chemical poisons and by desiccation.

The migration of spermatozoa. Normally the male fluid is ejaculated upon and about the cervix. Yet the spermatozoa may traverse the entire length of the genital tract by their own locomotory powers, and impregnation is therefore possible without introception of the male organ. Locomotion is accomplished by the vibratile motion of the tail. The rate of locomotion is about one inch in seven and a half minutes.

Place, time, and method of impregnation. Impregnation is by most authorities believed to take place in the outer portion of the Fallopian tube.   In the great majority of cases it occurs within a week after the cessation of a menstrual period.   The spermatozoön penetrates the ovum, its tail is absorbed, and its head forms the *male pronucleus*.   As a rule, a single ovum is fecundated by a single spermatozoön.   The male and female pronuclei unite to form the *vitelline nucleus* of the fecundated egg.

DEVELOPMENT OF THE IMPREGNATED OVUM.

The egg on leaving the ovary consists of the parts above described, and has a diameter of $\frac{1}{120}$ inch.   Cells of the membrana granulosa partially envelop it on its escape from the ovary.   In course of its passage through the oviduct it receives an albuminous envelope which supplies the first nutriment for its development.

Segmentation of the vitellus, or yolk, begins in the vitelline nucleus and continues until the whole yolk becomes a granulated mass.   Segmentation in the human subject probably does not occupy more than six or eight days.   By the time it is complete the ovum has a diameter of $\frac{1}{50}$ to $\frac{1}{25}$ inch, and has usually reached the cavity of the uterus.

The blastoderm.   Each granule forms a separate cell. A portion of these cells unite to form a continuous layer which lines the vitelline membrane.   This is the blastodermic membrane.   Cleavage takes place through the vitelline nucleus, its ultimate segments forming the nuclei of the cells of the blastoderm.

The blastodermic vesicle.   By the accumulation of fluid in the ovum it becomes converted into a vesicle known as the blastodermic vesicle.

Blastodermic layers. A second membrane is soon formed lining the first. The external blastodermic layer is called the *epiblast* or *ectoderm;* the internal, the *hypoblast* or *entoderm.* Between these two a third layer (the *mesoblast* or *mesoderm*) is subsequently developed.

From the epiblast are formed the epidermis, the cerebro-spinal axis, and the organs of special sense.

From the mesoblast are developed bone, muscle, connective tissue, the heart and bloodvessels and the genito-urinary organs.

From the hypoblast the epithelium of the alimentary and respiratory tracts and glands is formed.

The area germinativa or *embryonic spot.* There soon appears upon the blastoderm an opaque, oval spot, consisting of an aggregation of cleavage cells on the inner surface of the membrane. This is the area germinativa.

The area pellucida is a clear oval space, which appears a little later in the center of the area germinativa.

The primitive trace is a shallow groove or sulcus which now develops lengthwise through the center of the area pellucida. This is the first indication of embryonic structure and marks the place of the cerebro-spinal canal.

The dorsal laminæ are two longitudinal folds, which spring up on either side of the primitive trace. They arch over, and by the end of the first month unite to form the cerebro-spinal canal.

The ventral laminæ develop and unite in similar manner to form the cavity of the trunk.

The cephalic and caudal extremities of the embryo are developed out of folds which spring up at either end of the primitive trace.

The umbilical vesicle. By the folding-in of the ventral

plates the inner layer of the blastodermic vesicle becomes divided into two parts, the embryonic portion and the umbilical vesicle. The former is enclosed in the trunk cavity, the latter remains outside of it. The embryonic portion constantly increases in size, the umbilical vesicle diminishes till at the end of the sixth week it is no larger than a pea, and finally disappears. The umbilical vesicle contributes to the nourishment of the embryo pending the development of the vascular system which ultimately unites the mother and the fetus.

DEVELOPMENT OF THE EMBRYO AND FETUS.

End of First Month.—Ovum the size of a pigeon's egg, diameter three-fourths of an inch; length of embryo, one-third of an inch. First rudiments of fetal structure discernible. Heart, kidneys, liver, extremities, and the eyes, oral and anal orifices begin to be formed. Heart begins to beat third week. Spinal canal closed.

Second Month.—Ovum the size of a hen's egg, two and a half inches in diameter; length of embryo, one and a quarter inches; average weight, one drachm. Rudimentary vertebræ are present. The frontal unites with the superior maxillary processes. Centers of ossification are apparent in the inferior maxillary bones and the clavicle. The visceral arches are nearly closed. Eyes, nose, and ears begin to be developed. Rudiments of hands and feet appear, webbed. The umbilical cord is an inch or more in length. Sexual organs apparent.

Third Month.—Ovum size of a goose's egg, diameter four inches; embryo about three and a quarter inches in length; weight, one ounce. The product of conception now, for the first time, occupies the whole cavity of the uterus. The placenta is distinctly formed; the chorionic villi have atrophied over two-thirds the surface of the ovum. The umbilical cord begins to be twisted. The external parts of the embryo are well formed. Ossific centers appear in most of the bones. Rudimentary finger- and toe-nails appear. The cavities are completely closed. Sex is differentiated by the presence or absence of a uterus. Active movements begin during the latter part of this month.

Fourth Month.—Length of fetus, five inches; average weight, about three ounces. Ossification is well established in frontal and occipital bones. The sex is clearly defined. Lanugo is present. The placenta is complete.

Fifth Month.—Length of fetus, nine inches; average weight, nine and a half ounces. The cord is about one foot in length. The eyelids begin to open. Ossification begins in the ischium. Hair and nails begin to develop. First appearance of vernix caseosa. Heart sounds audible.

Sixth Month.—Length of fetus, twelve inches; weight, about twenty-three ounces. Ossification begins in the pubic bones.

Seventh Month.—Length of fetus, fourteen inches; average weight, two and a half pounds. The pupillary membrane begins to disappear. In boys the testes are in the scrotum—at least the left one. Ossification begins in the astragalus. The fetus is viable, but viability is feeble.

Eighth Month.—Length of fetus, sixteen inches; average weight, three and a half pounds. The nails are completely developed, but not projecting beyond the finger tips. Ossification begins in lower epiphysis of the femur. A child born at this period is viable. Lanugo begins to disappear from the face.

Ninth Month.—Length of fetus, seventeen inches; diameters of the head, a half to two-thirds of an inch less than at term; average weight, about four pounds. Lanugo begins to disappear from the body.

Tenth Month.—*Signs of maturity.* Measurements: length, eighteen to twenty inches; sub-occipito bregmatic circumference, thirteen inches; length of foot, three and one-eighth inches. Weight, seven to seven and a quarter pounds. Eyes usually open. Face and body plump. Suckles. Cries lustily. Lanugo has almost entirely disappeared from the body. Vernix caseosa is present, chiefly on the back and on the flexor surfaces of the limbs. The finger-nails project beyond the finger tips, the toe-nails to the end of the bed of the nail. The cartilages of the ear and the nose are firm. Cranial bones hard, sutures and fontanelles small. Centres of ossification well developed in the lower epiphysis of the femur and in the astragalus, beginning in the upper epiphysis of the tibia and the cuboid bone.

FETAL CIRCULATION.

The peculiarity of the fetal circulation is due mainly to the fact that pulmonary respiration is in abeyance during intra-uterine life, the respiratory blood changes being effected in the placenta. Only enough blood goes to the lungs for their nutrition. The blood passes from the placenta through the umbilical vein. A part goes direct to the ascending cava by the ductus venosus, and a part reaches it through the liver and hepatic vein. With the blood from the lower extremities it then passes to the right auricle, and is deflected by the Eustachian valve through the foramen ovale into the left auricle, thence into the left ventricle and into the aorta. The larger part goes to the upper extremities and the head. Thence it returns by the descending cava to the right auricle, passes to the right ventricle, a very small portion going to the lungs by the pulmonary artery, the main portion entering the aorta through the ductus arteriosus; a small part of this mixed blood goes to the lower extremities, the greater part returning to the placenta by the hypogastric arteries.

The umbilical arteries are closed within two days after birth. The umbilical vein and ductus venosus, generally in six to seven days. The ductus arteriosus, within nine to fifteen days.

THE DECIDUÆ.

a. The decidua vera is the hypertrophied mucous membrane of the uterus. It increases in thickness tenfold in the first month, to two-fifths of an inch.

The decidua serotina or *placental decidua* is the part of the decidua vera which underlies the ovum, where the placenta is subsequently to be developed.

b. The decidua circumflexa or *reflexa* or *epichorial decidua*. When the ovum becomes imbedded in the folds of the uterine mucosa, that structure grows up around it. This reflected envelope is the decidua reflexa or circumflexa. It grows with the ovum and comes in contact with the decidua vera by the end of the third month, blending with it. The entire cavity of the uterus is from this time occupied by the ovum and its membranes. Except at the placenta the deciduæ undergo atrophy, and are reduced to a single thin membrane by the end of the third month; the decidua reflexa wholly disappears by the end of the seventh.

THE AMNION.

The amnion is the innermost of the fetal envelopes. As soon as the embryo begins to take shape, a fold of the epiblast springs up entirely around the edges of the embryo. This membranous ridge develops till its edges meet over the back of the embryo. The surfaces brought in contact fuse together. The neck of the pouch thus formed is subsequently absorbed. The pouch itself is the amnion. In it is included the embryo. It is gradually expanded by accumulation of fluid in its cavity— the liquor amnii. Ultimately a sheath of the amnion invests the umbilical cord and is continuous with the fetal epidermis at the umbilicus.

The outer blastodermic layer, which is termed the false amnion, recedes to the vitelline membrane.

Structure. It consists of a layer of connective tissue and one of endothelial cells. It has neither vessels nor lymphatics.

Its function is to secrete and hold the liquor amnii.

The liquor amnii is a clear alkaline fluid. During the first half of pregnancy it is contributed directly from the body of the fetus and from certain capillary blood-vessels of the placenta immediately beneath the amnion.

In the later months it consists partly of fetal urine. Its constituents at term are water, a trace of albumin, saline matter, urea, epithelium. The amount at term is one to two pints. Its specific gravity is 1007.

Uses. During gestation it permits active fetal movements and serves to protect uterus and fetus. During parturition it also helps to dilate the cervix.

THE ALLANTOIS AND CHORION.

The allantois is a diverticulum which is pushed out from the posterior portion of the entoderm or intestinal canal at about the time when the amniotic folds are developed. It is projected to the external envelope of the ovum, which now consists of the vitelline membrane and the false amnion united in one. It expands till, by the end of the third week, it completely envelops the embryo and its investing amnion, and lines the external envelope of the ovum as a flattened sac.

Its office is to project vessels from the embryo to that portion of the outer envelope where the placenta is to be developed, the chorion frondosum.

The pedicle of the allantois ultimately dwindles to a mere cord known as the umbilical communication, the rudimentary umbilical cord.

The chorion is formed by fusion of the allantois with the vitelline membrane and the false amnion. The space which for a time exists between the amnion and the chorion is occupied with a gelatinous material, in which the umbilical vesicle lies enveloped.

Union of the fetal envelopes. The amniotic sac expands till it comes in contact with the chorion and unites with it at the end of the second month. The coverings of the ovum, then, from within outward, are the amnion, the chorion, the decidua reflexa, the decidua vera. They are practically a single membrane after the third month. The ovum loosens its hold upon the uterus at term by a process of fatty degeneration in the decidual layer.

Chorial villi. Soon after the fixation of the egg the surface of the chorion becomes everywhere covered with transparent villi. The villi penetrate the substance of the decidua, from which they derive nutrient material for the sustenance of the ovum. They are at first single, but with the growth of the ovum they elongate and become compound. The outer surface of the globular ovum thus becomes everywhere " shaggy."

Bloodvessels of the villi. The villi at first are not vascular, but they soon receive vessels from the allantois. The capillaries of the chorial villi enter the stem of the villus, follow the subdivisions, go to the end of each rootlet, turn, forming loops, and go back to empty into the venous trunks of the chorion. The chorial villi are somewhat similar to those of the intestines in structure and in function.

Chorion læve. Toward the end of the second month the chorion begins to become bald everywhere except over the portion about the insertion of the fetal blood-vessels. Over two-thirds the surface of the chorion the villosities gradually atrophy, until by the end of the second month this part of the chorion becomes smooth, the *chorion læve.*

Chorion frondosum. Over the remaining third of the chorial surface the villosities develop more rapidly than before, till this portion presents a thick, spongy mass of villosities, the *chorion frondosum*. They enter into the formation of the placenta. In the placental portion of the chorion the development of the vessels keeps pace with the growth of the villosities; elsewhere the capillaries shrink with their villi.

After the formation of the placenta the non-placental portion of the chorion serves only for protection.

THE PLACENTA.

The placenta when fully formed is a spongy mass of lenticular shape, measuring from seven to eight inches in diameter, and about one inch in thickness at the insertion of the cord. Its average weight is one pound.

The fetal surface is a smooth surface of amniotic membrane. The umbilical cord is attached generally at or near its centre.

The maternal surface is rough and divided into irregular lobes or cotyledons from one-half to one and a half inch in diameter, sixteen to twenty in number. These lobes are separated by membranous partitions which penetrate the substance of the placenta at their borders, as far as the fetal surface. The maternal surface is covered with the outer layer of the decidua serotina. The placenta after separation is studded with small openings, mouths of the veins and of the curling arteries of Hunter.

The placental seat is in the upper segment of the uterus, on the anterior or posterior wall with about equal frequency; rarely the insertion is lateral.

Development. The formation of the placenta begins in the second month of gestation. The limits of the placenta are distinctly defined at the end of the third; its characteristic form and structure are complete at about the end of the fourth month.

The chorionic villi penetrate the inter-glandular portion of the mucosa, and ramify to form dendritic tufts. The walls of the crypts into which the villi and their branches dip are lined with epithelium and are very vascular. The capillaries around the crypts enlarge and inosculate till every loop of the fetal villi is enveloped by a meshwork of dilated maternal capillaries. The maternal capillaries enlarge, obliterate the interspaces, and coalesce into great lakes of blood. These lakes communicate freely with the uterine sinuses.

Structure. The placenta consists essentially of blood-vessels. The vascular fetal tufts, sixteen to twenty in number, are suspended in lakes of maternal blood. The lakes are supplied by the curling arteries of the uterus. The maternal blood returns from the spaces between the fetal tufts, by the coronary vein upon the margin of the placenta and by sinuses situated in the septa between the cotyledons. The fetal and maternal circulations do not communicate directly.

Function. The placenta is at once the nutritive, respiratory, and excretory organ of the fetus. The interchange between the fetal and maternal blood takes place by osmose.

The Umbilical Cord is the pedicle which connects the foetus with the placenta. The fetal insertion of the cord is at the umbilicus. The placental insertion is generally central.

The average length of the cord is about twenty inches, varying from seven, or less, to sixty inches. Its size is

about that of the little finger of the adult.   The tensile
strength, at term, is from five to twelve pounds.

Structure.   At about the end of the first month it
becomes invested with a process of the amnion which is
continuous with the fetal epidermis.   It contains the
umbilical vessels imbedded in a jelly-like connective
tissue, the gelatin of Wharton.

Vessels.   It has, primarily, two arteries and two veins;
subsequently one of the veins disappears.   Exceptionally there is but one artery.   The walls of the arteries
are only a little thicker than those of the veins.   The
cord is usually twisted, the vein being wound around
the arteries.   It has neither nerves nor lymphatics.

EFFECTS OF PREGNANCY ON THE MATERNAL ORGANISM.

Changes in the Uterus.—The first effects of pregnancy are noted in the uterus.   It changes in size, shape,
and structure.

Size.   Growth begins immediately with the fixation
of the ovum, and is continuous with its growth.   Its
development during the first two months is chiefly in
the lateral and antero-posterior directions.   In the later
months the enlargement is partly by dilatation.   The
thickness of the uterine walls at term is between one-
sixth and one-quarter inch.   The internal surface is increased between conception and full term from five or six
to three hundred and fifty square inches.   The capacity
of the uterus is increased from one to four hundred
cubic inches—five hundred and nineteen times.   The
weight increases from one or one and a half ounce in

the pre-gravid state, to two or two and a half pounds
at term.

Approximate Measurements of the Gravid Uterus.

Stage of Gestation.	Total Length.	Width.
12 weeks.	5 inches.	4 inches.
16 "	6 "	5 "
20 "	7 "	6 "
24 "	8½ "	6½ "
28 "	10 "	7 "
32 "	11½ "	8 "
36 "	13 "	9 "
40 "	14 "	10 "

Shape. In the first trimester the shape of the uterus
is pyriform; in the second, the corpus uteri is a flattened
spheroid; in the last, it is generally ovoid, the fundal
being the larger end.

Structure. The muscular fiber grows seven to eleven
times in length, and two to five in thickness; there is
probably, also, some hyperplasia of muscular tissue. At
the os internum there is a preponderance of circular
fibers in all the layers.

The arteries increase in length and caliber. The
uterine plexus of veins develops into a system of enor-
mous sinuses in the middle coat of the muscularis, and
in the inner coat beneath the placental attachment. The
lymph tubes become hypertrophied to the size of a goose-
quill, and the lymph spaces enlarged. Hypertrophy of
the nervous structures probably keeps pace with the
general development of the uterus.

Changes in the Cervix Uteri.

Size. The apparent shortening of the cervix during
gestation is due partly to softening, and partly to swelling of the vaginal mucous membrane and loose cellular
tissue around the cervix at the vaginal junction. The

cervical enlargement is due in part to hypertrophy, but mainly to loosening of its structure from serous infiltration, and is progressive to about the eighth month.

Structure. Softening extends from the lower border upward, and involves the entire cervix by the end of the eighth month. By this time the cervical canal has become sufficiently patulous in multiparæ to admit the finger, and the head of the child may generally be touched through the cervix.

Changes in other Pelvic Structures.—The peritoneal covering of the uterus is developed by tissue growth, *pari passu* with the development of the organ itself.

The broad ligaments accommodate themselves to the growth of the uterus partly by the unfolding of their layers and partly by hypertrophy and hyperplasia of tissue elements.

The ovaries and Fallopian tubes lie in contact with the uterus by the time it rises out of the pelvis.

The vagina becomes hypertrophied during pregnancy. It increases in the width and length of its walls and in vascularity.

General Changes.—According to most authorities, hypertrophy of the left ventricle of the heart during gestation is physiological, and is designed to meet the increased resistance in the systemic circulation introduced by the new vascular arrangements of the uterus and placenta. The cardiac movements are slightly accelerated.

The thyroid gland becomes hypertrophied during pregnancy, and a certain degree of enlargement remains permanent.

The total volume of the blood is increased in the latter half of pregnancy. The proportion of water is little, if at all, greater.

In most women there is a marked increase in the irritability of the nervous system during gestation. Psychical disturbances, neuralgias, and other nervous affections are common.

Frequently a considerable gain in the weight of the body occurs in the later months, due mainly to increased deposit of fat.

SIGNS OF PREGNANCY.

A. HISTORY.

Suppression of menses. In the absence of other appreciable causes, and in a woman of previously regular menstrual. habit, the arrest of the catamenia affords strong presumptive evidence of pregnancy. Other possible causes of the absence of menstruation are anemia, tuberculosis, syphilis, chronic nephritis, exposure to cold, change of climate, tardy menstruation, the menopause, and emotional causes.

This sign is not always available for diagnosis. Conception may take place in the absence of menstruation during lactation, or before the menstrual function is established. Again, periodical hemorrhages simulating menstruation sometimes occur in the early months of gestation. The bleeding in such cases usually proceeds from lesions of the cervix, chronic decidual endometritis, or placenta previa, and its occurrence at the end of the menstrual month is explained by the influence of the menstrual molimen. Usually it may be distinguished from true menstruation by the difference in the amount and duration of the flow.

Nausea is present for a longer or shorter period in the great majority of pregnancies. Generally it begins about the end of the first month and ceases by the end of the third, when the uterus rises out of the lesser pelvis, often earlier.

Usually it is a morning sickness. In exceptional cases it is more or less persistent throughout the day. Pathological causes must be excluded.

Ptyalism is a frequent accompaniment of the nausea of pregnancy. It is generally attended with excessive secretion of mucus in the mouth and throat. Marked salivation is exceptional.

Certain mammary and abdominal signs may be elicited from the history, such as enlargement, sense of weight, fulness and tenderness of the breasts, enlargement and pigmentation of the abdomen, and quickening.

B. PHYSICAL SIGNS.

I. *Mammary Changes.*

a. Increased size and fulness of the glands. The milk-glands are developed by growth of the acini, swelling of the connective tissue and by interlobular deposit of fat. Enlargement of the gland may be distinguished from overlying fat by palpation. The gland is readily identified by its nodular border. Frequently the fulness and firmness are not well marked after mid-pregnancy. Rarely there is no notable enlargement during the entire period of gestation.

b. Primary areola. The areolæ become pigmented, elevated, edematous. The depth of pigmentation varies with the complexion of the patient. It is faintly developed in blondes, marked in brunettes, and nearly black in the negress. Frequently it shades into the color of the skin at the upper and outer aspect of the areola at the end of the second month. The areola acquires a soft velvety feel, and is slightly lifted above the level of the surrounding skin.

c. *Montgomery's follicles.* The sebaceous follicles of the areola, ten to twenty in number, become hypertrophied during pregnancy. They appear as papular elevations over the surface of the areola. They are best displayed by putting the skin gently upon the stretch.

d. *Enlargement of veins.* Everywhere over the breasts the veins become more prominent. By slightly stretching the skin, they may be seen coursing across the areola.

e. *Milk secretion.* A milky serum may be pressed from the nipples after the third month. In women who have never borne children its presence affords valuable evidence of pregnancy. Yet milk secretion is possible in virgins, and even in males. The sign has no value after the first pregnancy, since milk may generally be found in the breast of parous women.

In examining for this sign, the manipulation should begin over the duct sinuses at the base of the nipple.

f. *Secondary areola.* The secondary areola is a faintly pigmented zone skirting the primary areola. It is characterized by the presence of one or more rows of feebly marked circular spots immediately about the primary areola. The markings correspond to non-pigmented sebaceous follicles. The secondary areola is diagnostic when well made out.

Date of appearance. With two exceptions all the mammary signs may be looked for by the end of the second month. Milk is present at the third, and the secondary areola appears at the fifth. The group of mammary signs is rarely complete.

Diagnostic value. In primiparæ the mammary changes frequently afford sufficient evidence for at least a pre-sumptive diagnosis of pregnancy. In women who have borne children they are not to be relied on, since many of them are not effaced after the first childbirth. Pathological conditions of the sexual organs must be excluded; breast changes similar to those of pregnancy may result from pelvic disease.

II. *Abdominal Signs.*

1. Inspection.

a. Flattening. In the second month the abdomen is slightly flattened; the uterus, during this period, sinks lower in the pelvis, and the hypogastrium is a little less prominent in consequence.

b. Enlargement begins about the end of the third month, when the uterus begins to rise out of the lesser pelvis, thereafter increasing with the growth of the uterus till the middle of the ninth month. Within two weeks or more before term the uterus usually sinks deeper in the pelvis, the fundus falls farther forward, and the upper abdomen becomes perceptibly smaller.

c. Pigmentation. Pigmentation of the abdomen is most frequently limited to a narrow band about one-eighth inch in width, extending from the pubes to the umbilicus, sometimes to the ensiform. It appears about the end of the second month. Pigmentation of the ab-domen, like that of the breast, varies in depth and area with the complexion of the patient. In brunettes a dark circle develops around the umbilicus, and pigmented patches occur over other parts of the abdomen. In blondes there is, sometimes, an entire absence of pigmentary changes. Similar deposits of pigment may occur in other conditions than pregnancy.

d. Umbilical changes. The umbilicus becomes retracted in the first three months, is protruded in the last two or three.

e. Lineæ albicantes, or *striæ gravidarum.* The lineæ albicantes are irregular whitish, pinkish, or bluish, lines developed over the lower half of the abdomen during the later months of pregnancy. They are due chiefly to partial atrophy from tension, and appear at about the sixth month. Once formed, they remain more or less permanent. Abdominal distention from other causes than pregnancy may give rise to similar markings.

2. **Palpation.**

a. Size of the tumor. The fundus uteri lies in the plain of the pelvic brim at the third month, is near the level of the umbilicus at the sixth, and touches the ensiform cartilage at the thirty-eighth week. For measurements of the uterus at different stages of gestation, see Table, p. 49.

b. Character of the tumor. The gravid uterus is smooth, symmetrical, pyriform or ovoid, fluid. In the last trimester, and even earlier, fetal parts may be detected by palpation.

c. Intermittent contractions of the uterus may be detected by the fourth month by palpation over the abdomen ; at an earlier period by the bimanual. They recur at intervals of five to ten minutes; may be obtained immediately by applying the hand cold, or by gentle friction over the tumor. They are not interrupted by the death of the fetus. Hematometra, hydrometra, distended bladder, and soft fibroids must be excluded.

Similar contractions, it is claimed, occur in the non-gravid uterus.

d. Active fetal movements.

As an objective sign, active movements of the fetus afford positive evidence of pregnancy. It is available by abdominal palpation about the fourth month. It is most promptly obtained by applying the hand cold to the abdomen, or by first tossing the fetus from side to side. Muscular movements of the fetus begin about the tenth week, and may frequently be detected by the bimanual examination as early as the twelfth. Detection of fetal movements is difficult or impossible in hydramnios, and in certain other conditions. They may be absent for a time, even in ordinary cases.

As a subjective sign, the movements of the fetus are not wholly reliable. They may be simulated by intestinal flatus, spasmodic contractions of the abdominal muscles, or other causes. The first sensation of fetal movements, as felt by the mother, is termed *quickening.* The period of quickening is usually the fourth month. It varies, however, from the twelfth to the twentieth week. Rarely, the movements are not perceived by the mother during the entire period of pregnancy.

e. Passive fetal movements; external ballottement. External ballottement is practised by placing the hands well apart upon the abdominal walls, with their palmar surfaces facing each other, and tossing the fetus between the hands. Pathological growths floating in fluid must be excluded.

3. Auscultation.

a. The funic souffle is a bruit synchronous with the fetal pulse. It is heard in but a small proportion of cases, and only in the later months. The cause of the bruit is pressure on the cord, impeding the blood current.

b. The uterine souffle is a soft blowing murmur synchronous with the maternal pulse. It is best heard over the lateral borders of the uterus, especially the left, since owing to the usual right torsion of the gravid uterus the left border is most accessible. It is generally present after the fourth month; may be detected earlier by pressing the stethoscope deeply down over the side of the uterus. The sound has its origin in the ascending uterine arteries and their branches, not in the placental sinuses, as formerly believed. It may be heard after the delivery of the placenta. Uterine fibroids, chronic metritis, and even ovarian cysts, may give rise to a similar souffle.

c. The choc fetal is the shock of the fetal movements as perceived by the ear on abdominal auscultation over the uterus. It resembles the effect produced by lightly percussing one hand held against the ear, with a finger of the other hand. The *bruit de choc fetal* is a bruit that immediately precedes the *choc fetal,* and is due to the displacement of liquor amnii by the fetal movements.

d. The fetal heart sounds become audible by abdominal auscultation at the fourth or fifth month. By vaginal stethoscopy they may be heard at the twelfth week. The character of the heart tones does not differ essentially from those of the newborn infant, heard through several thicknesses of clothing. The rate is about double that of the maternal pulse—120 to 150 per minute. They are audible over an area of about three inches in diameter, frequently more. The point of maximum intensity is the *focus of auscultation.* Usually this point lies over the lower angle of the left scapula of the fetus. Rarely, there may be two foci, even in single fetation. The second focus is explained by conduction through some remote point of fetal contact with the uterine walls. The heart sounds may be temporarily inaudible, owing to dorso-posterior position, hydramnios, or other causes. Their persistent absence may generally be taken as evidence of fetal death.

Method of examining. Patient in the dorsal decubitus, in a still room ; auscultate by the mediate or immediate method, with or without the stethoscope. Listen over the probable or ascertained location of the left fetal scapula. Failing there, search the whole surface of the tumor. Press the abdominal walls well against the tumor : a continuous solid medium helps conduction. In dorso-anterior positions, pressing the breech downward in the axis of the fetus helps by thrusting the dorsum forward. Failing, repeat the attempt at intervals of several hours or days.

A rhythmical succession of sounds of characteristic quality and rhythm, of about double the maternal pulse-rate, and which can be counted, establishes the diagnosis of pregnancy.

III. *Pelvic Signs.*

a. Purplish color of the vagina. The vagina assumes a purplish hue, which varies greatly in depth in different cases and in the same case at different stages of gestation. Usually a venous hue is faintly developed in course of the second month. It is most constantly found in the anterior wall immediately below the meatus urethræ. The cause of the deepening color is hypertrophy of the corpus cavernosum of the vestibule, and of the vaginal venous plexuses.   It is to be found in about five-sixths of all cases at the end of the third month.   Pathological congestion must be excluded.

Purplish color of the cervix. A similar change of color in the vaginal portion of the cervix may be observed almost from the first month after conception.

b. Softening of the cervix is usually perceptible to the touch by the sixth week.   At this early stage of gestation the softened portion is a thin layer at the lower border of the cervix ; it presents the feel of a thin velvety stratum covering the firm body of the vaginal portion.   As pregnancy advances the softening progresses from below upward, and involves the entire cervix by the end of the eighth month.   The cervical canal becomes more patulous as the cervix softens.   These changes are not always clearly defined in the early months.   Softening from pathological causes lacks the progressive character which belongs to pregnancy.

c. Changes in the uterine tumor. The most reliable evidences of pregnancy in the second and third months are the changes in the size, shape, and consistence of the uterus as detected by the bimanual examination.   The body of the uterus grows with the growing ovum, takes on a globular shape, and a soft, elastic feel.   These changes are well marked by the sixth week and may often be recognized at an earlier period.

The enlargement of chronic metritis or of subinvolution is distinguished from pregnancy by greater density, absence of growth, and by the history.

An anteflexed and hyperemic uterus may simulate

the shape and consistence of the gravid tumor, but it, too, is distinguished by the absence of growth.

A uterus containing a soft submucous fibroid may usually be differentiated by the history and by the rate of enlargement.

Hematometra and hydrometra present the characters of a tense cyst.

Hegar's sign. One of the most striking changes in the uterus at this period of gestation is the compressibility of the isthmus uteri, which has become known as Hegar's sign. It is particularly marked in the median portion, which in the non-gravid state is the most dense.

Method. The uterus is depressed by the external hand, or drawn down by a volsella. The thumb of the other hand is passed into the vagina and pressed against the lower uterine segment at its junction with the cervix. A finger of the same hand is carried into the rectum to a point just above the utero-sacral cul-de-sac. The tissues between the thumb and finger may be compressed almost to the thinness of a visiting-card. Thinning under pressure to less than a half-centimeter (0.2 inch) is diagnostic of pregnancy. No pathological condition simulates it.

Examination may be facilitated by the use of an anesthetic, and by distending the lower end of the rectum with water.

The compressibility of the isthmus may be made out by the use of the index finger of one hand in the anterior and of the other in the posterior vaginal fornix, the uterus being drawn down with a volsella; usually, by the ordinary method of bimanual exploration.

d. Pulsation of the uterine artery is perceptible by the touch through the vagina at one side of the cervix from the first months of pregnancy. Pathological causes must be excluded.

e. The temperature of the cervix ranges from $\frac{1}{2}°$ to $\frac{3}{4}°$ F. above that of the vagina.

f. Internal ballottement; passive fetal movements. Internal ballottement may be practised in the fifth and sixth months. Earlier the fetus is too light, later its mobility is too limited to permit of ballottement.

Method. The patient is placed in the semi-recumbent posture, with the bladder empty and the clothing loose. Two fingers in the vagina are placed against the anterior uterine wall above the cervix, while the outer hand steadies the fundus. The fetus tossed upward falls again, and repercusses the finger.

Distinguish from—

    Anteflexed uterus;

    Pedunculated tumor of the ovary or uterus;

    Internal projections of large cysts;

    Floating kidney;

    Stone in the bladder;

    Pulsations of the uterine artery.

Ballottement may fail from—

    Scanty liquor amnii;

    Abdominal presentation;

    Placenta previa;

    Multiple fetus, or

    Other causes.

SUMMARY OF DIAGNOSTIC SIGNS.

Mammary signs collectively, in first pregnancies ;
Intermittent uterine contractions ;
Detection of fetal parts ;
Active fetal movements ;
Changes in the uterine tumor, including
    Hegar's sign ;
Internal ballottement ;
Fetal heart sounds.

DIFFERENTIAL DIAGNOSIS.

Abdominal enlargement from other causes than pregnancy is distinguished from it by the absence of most of the signs of pregnancy, particularly those which pertain to the uterus. The non-gravid tumors of the abdomen also present certain characteristics of their own, by which they may usually be differentiated from gestation.

Fat in the abdominal walls may be lifted in folds and moved about over the underlying muscles.

A *phantom tumor* disappears under anesthesia.

Tympanites generally subsides in the morning, its percussion note is resonant, and palpation is negative. The abdominal walls can be gently pressed backward against the vertebral column.

In ascites the abdomen is flattened at the umbilicus when the patient lies in the dorsal decubitus, and the percussion note is tympanitic at the summit of the tumor, except in rare cases, where the mesentery is too short to permit flotation of the intestines to the surface of the fluid.

A fluid wave may be obtained through all parts of the

tumor within the limits of the fluid. (In pregnancy the wave is interrupted by the fetal mass.) The fluid level changes in different postures of the patient.

Tumors of other viscera may be traced to the normal situation of those viscera, and the uterus is easily differentiated from the tumor.

An ovarian cystoma is characterized by more marked fluctuation, usually, than the tumor of pregnancy, and by the absence of fetal parts, of active fetal movements, and of the fetal heart. The uterus may be differentiated from the tumor, generally.

Uterine myomata of the submucous variety are distinguished from pregnancy by the history of hemorrhage and by their greater density usually.

In subperitoneal myomata the tumor is nodular—it lacks the symmetrical shape and smoothness of the gravid uterus.

In both varieties the growth is not so rapid as in gestation. Pregnancy, however, may coexist with myomata or other pelvic or abdominal neoplasms, and is then often difficult of recognition.

PLURAL PREGNANCY.

Twins are met with in about one case in eighty or ninety, triplets in one in six to eight hundred. Quadruple and even quintuple pregnancies are possible.

Multiple pregnancies border closely on the pathological. The viability of the children is less than in single fetation. The fetuses are generally undersized, and of unequal development. The death of one or both *in utero* is not an uncommon occurrence. Usually in twin pregnancy there is excess of liquor amnii. In two-thirds of the cases labor is premature.

Origin of multiple pregnancy. Multiple pregnancy may result from the simultaneous rupture of two or more Graafian follicles in the same or different ovaries, from two ovules in one follicle, or from a single ovule with a double germ.

Children from the same ovule are always of the same sex.

Arrangement of the membranes and placentas. In twin fetation from two ovules, there are two amnions, two chorions, and two placentas. The placentas, however, may fuse at their margins, each having an independent circulation.

In twin fetation from a single ovule with a double germ, there is a single chorion containing two amnions; the placenta is single. Rarely there are two fetuses in a common sac, owing to destruction of the amniotic septum.

Super-fecundation is a twin pregnancy resulting from separate acts of insemination by the same or different males.

DURATION OF PREGNANCY.

The duration of pregnancy is not definitely known, since it is impossible to fix the date of fecundation.

The average interval between the beginning of last menstruation and labor is two hundred and eighty days; more accurately, ten times the menstrual interval which is habitual with the patient.

The average period between the fruitful coitus and labor is two hundred and seventy-three days.

A variation of at least twenty days above or below these averages is believed to be possible within physiological limits.

RULES AND METHODS FOR PREDICTING THE DATE OF LABOR.

a. Naegele's rule. Count forward nine calendar months from the beginning of the last menstruation and add seven days. This is a ready method of computing the 280 days from the beginning of the last menstruation. For prognosticating the date of labor it is usually accurate within a week.

Reckoning from the date of quickening is not wholly reliable. The period of quickening is subject to variations, and the observations of the patient are always liable to error.

b. Mensuration of the uterus is not reliable for the purpose, since the amount of liquor amnii varies in different cases, and the size of the fetus is not always the same in different instances for the same period of gestation.

Situation of the fundus. The fundus uteri usually reaches the plane of the brim at the third month, the umbilicus at about the sixth, the ensiform at eight and a half months, and, after lightening, sinks to a somewhat lower level. But the accuracy of this method is vitiated by the same causes as the preceding, and still further by the fact that the umbilicus is not a fixed point.

c. Mensuration of the fetus. The length of the fetus is approximately double that of the fetal ovoid. The length of the fetal ovoid may be measured with approximate accuracy through the abdominal walls, or by placing one pole of a pelvimeter against the head through the vagina, and the other upon the abdomen over the breech. The rate of fetal development, however, is not uniform. Yet the measurements of the fetus, including the diameters of the head, afford fairly reliable data for predicting the date of labor.

Length of the fetus.

Sixth calendar month, 30 to 35 cm. (about 12 to 14 inches).

Seventh calendar month, 35 to 40 cm. (about 14 to 16 inches).

Eighth calendar month, 40 to 45 cm. (about 16 to 18 inches).

Ninth calendar month, 45 to 50 cm. (about 18 to 20 inches).

HYGIENE OF PREGNANCY.

The pregnant woman should place herself under the direction of her physician from the first months of pregnancy. She should be urged to advise him promptly of even slight departures from health, and particularly during the later months.

Important hygienic requirements are: Open-air exercise for one or two hours daily, with care to avoid exhaustion and violent muscular exertion; avoidance, if possible, of depressing emotions and all injurious mental influences; regularity of meals; proper quantity and quality of food; daily evacuations of the bowels; eight hours of sleep daily; pure air at all times; a tepid sponge bath twice weekly in winter, once daily in the summer months.

A vaginal injection of a pint of water at a temperature of 98° F., or of a weak borax solution—℥ss ad Oj —may be used once or twice daily in case of irritating leucorrheal secretions.

Clothing. Light flannel underwear should be worn at all seasons, and the outer clothing should be changed to suit climatic changes. Tight clothing must be avoided, especially about the breasts and abdomen, and all heavy garments should be suspended from the shoulders.

Care of the nipples. During the last two months of pregnancy it is a good practice to cleanse the nipples daily with a weak borax solution (℥ss ad Oj). They may be anointed with fresh cocoa-butter after cleansing, and if they are small or sunken the patient should make daily practice of drawing them out with the thumb and fingers.

The urine should be examined for albumin once a week during the last two months, oftener in case of any suspicion of nephritis. It is a wise precaution to make an occasional examination at earlier periods. The microscopic findings are of greater value for diagnostic purposes than the chemical tests alone.

Quantitative tests for urea afford the most exact evidence of the manner in which the kidneys are performing their functions.

Marital relations during pregnancy should be restricted, and especially at the menstrual dates. The violation of this rule is a frequent cause of abortion and of premature labor.

PATHOLOGY OF PREGNANCY.

DISEASES OF THE DECIDUÆ.

Acute endometritis may develop in the course of acute febrile diseases. It is frequently attended with hemorrhage, and often results in abortion.

Chronic diffuse endometritis. The causation is not well understood. The anatomical changes in the decidua are chiefly hypertrophic. It sometimes gives rise to abortion.

Polypoid endometritis is characterized by polypoid growths in addition to the lesions of simple diffuse endometritis; it terminates in abortion if the chorionic villi become involved.

Cystic endometritis is distinguished by the formation of retention cysts by obstruction of the gland ducts through proliferation of inter-glandular connective tissue.

Catarrhal endometritis is attended with discharges of thin, watery mucus from the uterus—*hydrorrhœa gravidarum;* it is most common in the later months of gestation. The fluid collects sometimes between the chorion and the decidua, and is discharged in gushes.

ANOMALIES OF THE AMNION AND THE LIQUOR AMNII.

Oligo-hydramnios. The normal quantity of amnial liquor at term is about two pints. Oligo-hydramnios is a deficiency of liquor amnii. Deficiency of the amnial liquor may be attended with adhesions between the amnion and the fetus and with the formation of amniotic bands. Intra-uterine amputation of fetal extremities and grave faults of fetal development sometimes result from these amniotic bands.

Hydramnios may be defined as an excess of liquor amnii over four pints.

Causes. Among the causes assigned are abnormal persistence of the vasa propria (a capillary network of the placental portion of the chorion immediately underlying the amnion, normally present in the early months of gestation), excessive urinary secretion by the fetus, excessive excretion of the fetal skin, amniotitis, transudation from the maternal blood, deficient resorption of liquor amnii.

Diagnosis. The principal physical signs are excessive size and permanent tension of the tumor, supra-pubic edema, preternatural mobility of the fetus. Distinguish from ascites, ovarian cysts, twins, by palpation and auscultation of the tumor and by the history.

Prognosis. Unfavorable for the child, owing to premature birth, dropsical affections, malformations and malpresentations, which are common in hydramnios. Fetal mortality, 25 per cent. The prognosis for the mother is generally good.

Treatment. In case of alarming symptoms from over-distention, puncture the membranes, guarding against syncope from too rapid escape of the liquor amnii. On birth of the child precautions may be needed to prevent post-partum hemorrhage.

DISEASE OF THE CHORION.

Vesicular mole, hydatidiform mole, may be defined as "an hypertrophy and myxomatous degeneration of the chorial villi, attended with the formation of cysts." The cysts vary in size from that of a millet-seed to a grape—may reach the size of a hen's egg. They contain a watery fluid containing albumin and mucin. The degeneration begins in the very first weeks of gestation.

Etiology. The etiology is obscure. The cause apparently resides wholly in the ovum. Endometritis, syphilis, and absence of allantoic vessels, commonly assigned as causes, probably have no part in the etiology.

Prognosis. Maternal mortality, 10 to 15 per cent., from hemorrhage, sepsis, or rupture of the uterus. Except in rare cases of partial degeneration the fetus invariably dies. The degenerated ovum may be retained for many months, usually is expelled before the sixth.

Diagnostic signs.

Signs of pregnancy ;

Abdominal enlargement out of proportion to the period of gestation—uterus too large the first three months, later too small ;

Absence of ballottement, of the fetal heart, fetal parts, and fetal movements ;

Uterus usually boggy ;

Sanguineous discharge ;

Discharge of cysts—rarely noted ;

Detection of the cysts by exploration of the uterine cavity.

Treatment. As a rule, empty the uterus   Dilate the

cervix and evacuate with the hand, cautiously, since the uterine wall is often extremely thin. Curette after considerable retraction has taken place. Wash out the uterus with a hot antiseptic douche. Swab the cavity with tincture of iodine. Give ergot, if required, to make the uterus contract.

ANOMALIES OF THE PLACENTA.

Placenta membranacea. A placenta membranacea is a broad, thin placenta, with persistence and development of the villi over the entire surface of the chorion.

Placenta previa. The placenta is previa when its attachment encroaches upon that zone of the uterus which is subject to dilatation during the first stage of labor.

Placenta succenturiata. This term applies to a wholly or partially independent placental cotyledon. The anomaly may be single or multiple.

Cysts of the placenta are of common occurrence. The cysts are small and seated beneath the amnion. They are probably developed from fetal villi.

Syphilis. The syphilitic placenta is larger and paler than normal. Syphilis of the placenta is always dangerous, frequently fatal to the fetus.

Edema may occur in hydramnios, occlusion of umbilical veins, or maternal anasarca.

Apoplexy. Extravasations of blood into the placenta may take place at one or several points. They are a frequent source of abortion. The causes are placentitis, nephritis, pelvic congestion, mechanical violence.

Myxomatous degeneration usually involves only a part of the placenta. (See Vesicular Mole, p. 71.)

Fatty degeneration may proceed from endometritis, placental hemorrhage, chronic inflammation of the placenta. Death of the fetus sometimes results.

Placentitis probably owes its origin to an endometritis existing at the time of conception. Abnormal adhesions of the placenta arise from this cause.

Calcareous degeneration is common and unimportant.

White infarcts are very frequently observed in the placenta. They are unimportant when small and few in number. When extensive they may cause death of the fetus. They originate in local degenerations of the decidua.

ANOMALIES OF THE UMBILICAL CORD.

Length. Too long a cord may predispose to prolapse, to knots or to coils about the fetus and possible obstruction in the funic vessels. Too short a cord may lead to premature separation of the placenta during labor.

Excessive torsion may cause occlusion of vessels; yet in most cases this condition is developed only after the death of the fetus.

Knots occur rarely. They are seldom tight enough to endanger the fetus.

Hernia. Hernial protrusion of abdominal viscera may take place into the cord. It is usually accompanied with other errors of fetal development.

Cysts are frequently noted in the sheath of the cord. They are caused by liquefaction of mucous tissue or by blood extravasations.

Coils about the fetus, especially the neck, are of common occurrence. Rarely is the circulation impeded either in the cord or the girdled member. Extensive coilings may give rise to the dangers of short cord.

The insertion may be eccentric, marginal, or velamentous. In the latter anomaly the vessels pass for some distance between the membranes to the edge of the placenta. As they are more or less separated and unprotected they are liable to be torn during labor.

PATHOLOGY OF THE FETUS.

Anomalies of Development.[1]

a. Hemiteria. Literally, half monstrosity. This class includes dwarfs and giants, microcephalus, sternal fissure, spina bifida, encephalocele and other hernial protrusions; club-foot, patulous foramen ovale, imperforate rectum, vagina, esophagus, webbed fingers or toes, harelip, cleft palate, epispadias, hypospadias, supernumerary fingers or toes, etc.

b. Heterotaxia. Lateral transposition of viscera.

c. Hermaphrodism. Having the anatomical characteristics of both sexes.

d. Monstrosity.

1. Ectromelic monster. Having one or more aborted limbs.

2. Symelic monster. Having its lower limbs more or less completely united.

3. Celosomatic Monster. Having complete or partial eventration.

4. Exencephalic Monster. One in which the brain is malformed and protruding from the cranial cavity.

---

[1] See Norris's Syllabus.

Abortion.

Frequency.—Not far from 20 per cent. of all the pregnancies end in abortion. A large proportion of abortions occur at the end of the second month. This accident is comparatively infrequent after the third month.

Causes.

1. Death of the fetus from malformation, disease, mechanical violence, maternal toxemia, excessive anemia, morbid conditions of the chorion, the amnion, the cord, the decidua. The great majority of cases fall under this head.

2. Causes acting independently of the death of the fetus, such as:

Reflex irritation of the uterus;

Oxytocics;

Placenta previa;

Epileptiform convulsions from uremic or other causes;

Carbonic dioxide poisoning;

Placental apoplexies;

Pelvic adhesions;

Fibroids of the uterus;

Carcinoma of the uterus;

Misplacement of the uterus;

Overdistention from hydramnios or multiple pregnancy;

Direct interference;

Falls or blows;

Hyperemia of the pelvic organs from circulatory obstruction in the lungs or liver, cardiac disease,

violent muscular exertion, sexual excesses, etc.,
causing hemorrhage into the placenta ;

Atrophy or hypertrophy of the uterine mucosa.

Diagnosis.

Symptoms :

Hemorrhage ;

Pelvic tenesmus ;

Rhythmical uterine pains.

Physical signs:

Os internum expanding ;

Cervix dilating ;

Ovum protruding.

The physical signs make the diagnosis of inevitable abortion. Severe rhythmical pains with hemorrhage almost surely result in the expulsion of the ovum. Examine clots, breaking them up under water, for fetus or fetal appendages.

Prognosis.—In well-managed cases the prognosis is always good, and yet many deaths occur from mismanagement. The chief sources of danger are hemorrhage and septicemia. Hemorrhage contributes to the fatal issue, though rarely the immediate cause of death. The danger of sepsis is particularly imminent in incomplete abortion. The retention of necrotic material in the uterus is a serious menace to life. It is a common source of pelvic inflammation in cases which escape a fatal termination.

Treatment.

a. Prophylaxis in habitual abortion.

Syphilis in one or both parents, retroversion of the uterus, and endometritis are the most common causes of habitual abortion. Treat syphilis as in other cases, correct retroversion, and prevent recurrence by the use of a suitable pessary till after the third month. Endometritis is best treated by the curette in the interval between pregnancies. Guard against severe muscular exertion, mechanical violence, and the causes of pelvic congestion, especially at the menstrual dates. Enjoin rest in bed during the menstrual epochs and abstention from sexual intercourse till the critical period has passed.

b. Arrest of threatened abortion.

Maintain absolute rest in the recumbent position. The patient must not be permitted to rise for any purpose till all symptoms of abortion have subsided. Establish uterine rest by the use of opium, gr. ij, or its equivalent, p. r. n. Pil. ext. viburni prunifolii, gr. iv, q. 2 h , is valuable as an adjunct, even as a substitute for opium. Correct misplacement of the uterus. Exclude death of the fetus and vesicular degeneration of the chorion, in either of which conditions the uterus should be emptied.

c. Management of actual abortion. The chief objects of treatment are the prevention of—

　　Hemorrhage ;
　　Septicemia.

Agents for controlling hemorrhage are—

　　Rest ;
　　Vaginal tampon, simple or styptic ;
　　Evacuation of the uterus by the aseptic finger or curette.

Means for preventing or arresting sepsis are—

　　Avoidance of preventable lacerations and abrasions ;
　　The use of antiseptics ;
　　Early evacuation of the uterus.

1. *Expectant plan: Indications.* Ovum not detached, hemorrhage slight, absence of septic or putrefactive fluids.

Method. Usually no interference is practised except such as is required for cleanliness. An aseptic vaginal tampon may be used if required as a prophylactic against hemorrhage. This plan failing, after one or at most two days empty the uterus with curette and forceps—sooner for cause.

Method of tamponade. Sims' position and Sims' speculum. Material for tampon, sterilized cotton wool, used wet enough to pack firmly and in pledgets the size of a chicken's egg. Pack a row of pledgets in the fornix, around the cervix, and build up from this until the vagina is filled. Press the tampon away from the urethra and base of the bladder to prevent vesical irritation. Hold in place with a T-bandage. Sterilized gauze, in strips two and one-half inches in width and five yards in length, may be used instead of cotton wool. The simple aseptic tampon must be renewed every twelve hours. A tampon impregnated with zinc oxide may stand twenty-four hours. Mercurials must not be used in the tampon.

2. *Radical plan: Indications.* Cervix dilated, ovum detached or presenting or partially expelled, dangerous hemorrhage, sepsis present or imminent.

Manual method. An anesthetic may be used if required. The uterus is depressed and fixed with one hand over the abdomen, and the cavity emptied with one or two fingers of the other, aseptically. The manual method is impracticable, except the ovum is nearly or quite detached and the cervix well open, and is even then inferior to the instrumental.

Instrumental method. Anesthesia is generally necessary. The abdomen, thighs, vulva, and vagina should be thoroughly cleansed with a soap and water bath and washed with the antiseptic solution. The patient is placed in the Sims position and the cervix exposed by means of a Sims' speculum. The anterior lip of the cervix is caught and held gently forward with a volsella. The uterine cavity is douched with the antiseptic solution. The ovum is separated with a curette and removed with a pair of long straight dressing forceps having a joint about two and one-half inches from the distal end. Every part of the cavity is then thoroughly curetted and again douched with the antiseptic solution. The uterus may be swabbed with tincture of iodine if hemorrhage is not controlled by the curette. A flabby uterus after abortion calls for ergot.

Leave ten or twenty grains of iodoform, iodol, or aristol in the uterus. If the secundines were necrotic pack the uterine cavity lightly with a strip of iodoform gauze one or two inches in width. The gauze should be removed after twenty-four or thirty-six hours.

The presence of a perimetritis does not forbid interference. It makes it, on the contrary, the more imperative. Sepsis of the uterine cavity tends to perpetuate the peri uterine inflammation by maintaining the supply of septic material.

Incomplete Abortion.—Continuous or irregular hemorrhage after abortion, and sepsis or failure of involution should be regarded as probable evidence that portions of the ovum have been retained. In such cases the uterus should be disinfected, curetted, and lightly packed with iodoform gauze.

6

After-treatment of Abortion.—The patient should be kept in bed for a week or more, with strict attention to cleanliness. If the uterine cavity has been left empty and clean after abortion subsequent interference within the passages will not be required. Watch the temperature and the character of the genital discharge.

Premature Labor.

The causes of premature labor are essentially the same as those of abortion. The course and management do not differ essentially from those of labor at term.

HYPEREMESIS OF PREGNANCY.

Etiology.—The pernicious vomiting of pregnancy is frequently, to a greater or less extent, a neurosis. In many instances the reflex disorder is dependent in part upon some anatomical lesion of the pelvic organs, such as uterine displacement, detention of the uterus in the pelvis, by adhesions or otherwise, decidual endometritis, induration of the cervix, erosion or inflammation of the cervix, or perimetritis. Lesions of other than the pelvic organs, especially of the kidneys, may be present as complicating causes.

Prognosis.—The majority of cases recover by the third or fourth month, when the uterus rises out of the pelvis. The prognosis is grave in the worst forms.

Treatment.

a. Dietetic measures. Under this head may be mentioned breakfast in bed, followed by sleep; sherry wine or strong coffee before rising; a glass of cold carbonic acid water or Vichy, plain or containing an alkaline bromide, ʒj to the syphon, several times daily ; other dietetic measures as practised in ordinary vomiting; humor the appetite.

Rectal alimentation may tide the patient over a crisis when stomach-feeding is impossible. Beef blood, beef juice, Leube's meat solution, or predigested milk, ʒiv, q. 6 h., is a suitable food for the purpose. Small doses of opium may sometimes be added to the nutrient enemas with advantage. Wash out the rectum daily during rectal feeding.

b. General therapy. Among the more useful measures are : Rest in bed for several days ; cocaine, gr. ⅛ to ¼, three or four times daily, or hourly until three or four doses are taken ; cocaine spray to the pharynx, 1 per cent. solution ; chloral, gr. xx to xxx, in solution, by the rectum, repeated p. r. n., best given in milk ; the bromides in similar doses ; oxide of silver, gr. ¼ to ½, four times daily, on an empty stomach ; morphine, in doses of gr. ⅛ to ¼, hypodermatically, or endermatically at the epigastrium, especially if there be local tenderness ; strychninæ sulph., gr. $\frac{1}{40}$ to $\frac{1}{30}$, or tr. nuc. vom. ℥v, before meals, in chronic gastric catarrh ; calomel in full dose, gr. v to x, or in small repeated doses, gr. i, q. 1 h. ; oxalate of cerium, gr. x, q. 2 vel 4 h., in mild cases ; bismuth subnitrate in similar doses ; ether spray to the epigastrium at the beginning of each paroxysm ; ice-bag to the cervical vertebræ ; blister over the fourth or fifth dorsal vertebra ; inhalation of oxygen ; a mild faradic current through the stomach ; and other measures such as are used in the treatment of vomiting from other causes.

c. Local measures of value are the following : Pencilling cervical erosions with a twenty-grain solution of

nitrate of silver every second day; correction of malpositions of the uterus; abstention from sexual intercourse.

Galvanism. Anode against the cervix, cathode over
the lower dorsal vertebræ; current strength three to five
milliampères; sitting five minutes. Repeat twice daily.

Cocaine, 20 per cent. solution, applied freely upon
the portio vaginalis and within the cervix.

Combine with cocaine Copeman's method of dilatation of the cervix below the os internum. This treatment, however, may result in abortion.

The induction of abortion should be done as a last
resort, but should not be too long withheld. It must
be adopted, however, only with the concurrence of
council.

Method of Inducing Abortion.—Puncture of the
membranes, partial separation of the ovum with a clean
sound, or packing the cervix with iodoform gauze renewed every twelve to twenty-four hours.

The rapid method of evacuating the uterus with the
curette and uterine forceps is a good one in experienced
hands. The cervix is first dilated with an Ellinger
dilator till the curette passes readily. The operation
must be done under an anæsthetic.

PTYALISM.

Ptyalism is a reflex disorder like the nausea of pregnancy, with which it is usually associated. It is comparatively rare.

Treatment. Unsatisfactory. The following measures
may be tried: A saturated solution of potassium chlorate

used freely as a mouth-wash.   Atropinæ sulph., gr. $\frac{1}{64}$ once to three times daily, per os.   Bromides, gr. xxx to cxx, daily.   Salivation is usually most relieved by treatment which relieves the nausea.

ANEMIA.

Treatment.   Iron, pil. Blaud., 1 or 2, t. i. d.; arsenate of iron, gr. $\frac{1}{10}$, t. i. d.; albuminate of iron in liberal doses; citrate of iron subcutaneously, gr. j, in solution.   Combine with these agents the use of a generous diet.

VARICES OF THE LOWER EXTREMITIES

Are not uncommon in the later months of pregnancy. Treatment consists of support by bandage or elastic stockings.

PRURITUS VULVÆ.

Treatment.   Place the patient in the Sims position, retract the perineum with a Sims' speculum, and dust the vagina with subnitrate of bismuth.   Repeat once in one or two days.   Hot fomentations give temporary relief.   Applications of cocaine hydrochlorate are useful for the same purpose.   Exclude diabetes.

ECTOPIC GESTATION: EXTRA-UTERINE PREGNANCY.

In ectopic or extra-uterine pregnancy the fructified ovum lodges and begins development in the Fallopian tube.   Ectopic is, therefore, synonymous with tubal pregnancy.

Etiology.—Among the causes commonly assigned are partial obstruction of the tube, sacculation of the tube, crippled peristalsis, denudation of ciliated epithelium from old catarrhal inflammation and consequent loss of propelling power. The etiology, however, is unsettled.

Clinical Course.—According to the location of the fruit sac, two classes of cases may be distinguished : 1. Pregnancy in the free portion of the tube ; 2. Pregnancy in the intra-mural portion, or interstitial pregnancy.

A. *Pregnancy in the free portion of the tube.* Pregnancy in the free portion of the tube terminates almost invariably before the fourteenth week :

1. Rarely by death of the ovum without tubal rupture.

2. Usually by rupture of the tube.

Rupture may take place :

1. Into the peritoneum.

2. Into the broad ligament.

In either event more or less hemorrhage occurs from the tubal rent. In the latter the bleeding is, at the most, necessarily limited.

Intra-peritoneal rupture may terminate in :

1. Spontaneous arrest of hemorrhage and recovery in a small minority of cases.

2. Continuous hemorrhage and death.

Intra-ligamentous rupture may result in :

1. Immediate death of the ovum and blood collection—pelvic hematoma.

2. Continued development—intra-ligamentous pregnancy.

Intra-ligamentous pregnancy :

1. May go to term.   Spurious labor then occurs and the child dies.

2. May become intra-peritoneal by secondary rupture. The fetus rarely survives secondary rupture.

3. May die and be absorbed.

4. May die and suppurate.   A suppurating ovum may be discharged piecemeal through the abdominal wall, vagina, bladder, rectum ; may terminate in septicemia and death.

5. May die at or near term and be carried indefinitely, either with little or no alteration of structure, or may become encysted and calcified—lithopedion.

B. *Pregnancy in the intra-mural portion of the tube : tubo-uterine pregnancy : interstitial pregnancy.*   Terminates before the fifth month :

1. By intra-peritoneal rupture and fatal hemorrhage.

2. Possibly, in rare cases, by extrusion into the uterus.

DIAGNOSTIC SIGNS IN THE EARLY MONTHS.

1. *History :*
   Antecedent sterility ;
   Signs of pregnancy ;
   Pain ;
   Hemorrhage ;
   Decidual cast.
2. *Uterus :*
   Displaced, according to the size and situation of the
      fruit sac ;
   Enlarged, with rare exceptions ;

Empty ;
Cervix open.

3. *Tumor*—beside or behind the uterus :
Fluid ;
Tense ;
Tender ;
Pulsating ;
Rapidly growing.

Frequently there is a long period of sterility immediately preceding the pregnancy. The pain comes on
in paroxysms which are abrupt, usually violent, super-
vening upon apparent health, cramp-like in character
and generally referred to the seat of the fruit sac. The
final and more acute paroxysms are generally attended
with collapse and the signs of internal hemorrhage.

The genital hemorrhage is irregular in amount. It
occurs especially at the times of the painful paroxysms,
and a more or less copious discharge of blood commonly
attends the rupture of the fruit sac.

The decidual cast may come away entire or piecemeal.
It is distinguished by its histological characters from the
secundines of intra-uterine pregnancy and from the cast
of endometritis. Under the microscope it differs from
the former by the absence of fetal villosities ; from the
latter by the presence of decidual cells, which are round
or oval granular bodies, each containing a well-defined
nucleus or several nuclei, and having a diameter five to
fifteen times that of a red blood-corpuscle.

Exclude ovarian cyst, ovarian abscess, dermoid cyst,
intra-ligamentous cyst, fluid accumulations in the tube.

Distinction from pregnancy in the rudimentary horn
of a uterus unicornis is difficult or impossible ; but it is

practically unimportant, since the treatment called for in either condition is essentially the same.

DIAGNOSTIC SIGNS IN THE LATER MONTHS.

Fetal movements usually more distinct than in utero-gestation;

Fetal heart-tones are more intense;

Fetus more accessible to palpation;

· Ballottement in the fourth and fifth months;

Shrinkage of the tumor, usually after death of the fetus;

Uterus may be differentiated from the tumor.

Most reliable in the later months are the evidences of pregnancy with uterus but little developed and distinguishable from the tumor.

SIGNS OF RUPTURE.

Cramp-like pelvic pains, usually violent;

Genital hemorrhage;

Symptoms of acute internal hemorrhage with collapse;

Physical signs of pelvic hematocele or hematoma;

Signs of moderate peritonitis within two or three days after rupture;

In rupture with much internal hemorrhage the clinical picture is unmistakable. It is not so clear when the hemorrhage is small. Abortion and dysmenorrhea must be excluded.

Intra-peritoneal rupture is distinguished from extra-peritoneal by more hemorrhage usually, and by the physical signs of free fluid in the pelvic peritoneum. When the effusion of blood is limited by old adhesions, however, the condition cannot be distinguished from hemorrhage into the broad ligament.

Extra-peritoneal rupture is characterized by the evidence of a circumscribed and more or less solid tumor (blood-clot) in one broad ligament, as revealed by the vaginal touch. It may dissect up the peritoneum and burrow behind the uterus. Examination by the rectum, and, if necessary, under anesthesia, greatly aids the diagnosis.

Before opening the abdomen, either before or after rupture, the uterine cavity may be explored, remembering, however, that intra- and extra-uterine pregnancy may coexist.

TREATMENT BEFORE PRIMARY RUPTURE.

1. *Celiotomy and removal of the pregnant tube. Method.* The abdominal incision is made in the median line above the pubes and two or three inches in length. Adhesions are separated, the fruit sac, the ovary and the tube lifted, and the entire tube with the ovary tied off and amputated. The cut end of the tube is cauterized, all hemorrhage controlled, the peritoneum cleansed, and the abdomen closed.

2. *Feticide by electricity without puncture. Method.* Interrupted galvanic current; 120 interruptions per minute; one electrode on the abdomen over the tumor; one in the vagina or rectum beneath it. Strength of current, 20 to 50 milliampères. Sitting, ten minutes, repeated every twenty-four hours, till the tumor shrinks.

One sitting may suffice. A smooth galvanic current of 50 to 75 milliampères may be substituted in the absence of the interrupted.

The principal objection to electrical feticide is that the foreign body is left behind and may give rise to subsequent trouble.

TREATMENT AFTER RUPTURE INTO THE PERITONEUM.

Immediate celiotomy. Method, substantially as in section before rupture. Free blood in the abdomen may be detected by inspection through the peritoneum before it is cut or by passing a pipette through a minute opening in the peritoneum. The removal of the blood from the peritoneal cavity may be facilitated by irrigation with the normal salt solution (teaspoonful of salt to the quart of water), previously sterilized by boiling. In case of extreme anemia and collapse the patient should be prepared for operation by auto-transfusion (bandaging the extremities), hypodermatic injection of morphine, gr. $\frac{1}{4}$, or strychninæ sulphas, gr. $\frac{1}{30}$, or trinitrin, gr. $\frac{1}{100}$ to $\frac{1}{25}$.

Celiotomy being impracticable, the treatment consists in rest, with the use of sand-bags on the abdomen over the fruit sac.

TREATMENT AFTER RUPTURE INTO THE BROAD LIGAMENT.

Limited effusions of blood do not necessarily call for surgical interference. Should the cyst contents become septic or much hemorrhage occur, the abdomen should be opened. The edges of the incision into the sac should be stitched to the abdominal wound, the bleeding stopped, and the cavity drained.

TREATMENT AT OR NEAR TERM.

1. *Fetus living.* *Celiotomy.* The sac should be fixed to the abdominal walls by provisional suture before opening it. If practicable the placental site should be controlled by suture. The placenta and as much as possible of the sac may then be removed. The remainder of the sac should be closed and drained through the vagina or the abdominal wall. When the removal of the placenta is impracticable, the suture is left undisturbed and the sac drained. Its cavity must be kept clean by irrigation and other antiseptic measures. The placenta separates and may be removed within one or two weeks.

2. *Fetus dead.* Wait for obliteration of the placental vessels, if the patient is doing well, two or three months. Then by celiotomy remove the fetus and, if possible, the placenta and sac.

TREATMENT OF INTERSTITIAL PREGNANCY.

When diagnosis is possible, the pregnancy may sometimes be safely terminated by evacuating the fruit sac through the uterine cavity. On intra-peritoneal rupture, laparotomy is indicated as in pregnancy of the free portion of the tube, and supra-vaginal amputation of the uterus may be required.

PHYSIOLOGY OF LABOR.

THE MECHANICAL FACTORS OF LABOR.

THREE factors are concerned in the mechanical phenomena of labor—the powers, the passages, and the passenger.

I. THE EXPELLING POWERS.

The expelling powers are two.

1. Muscular action of the uterus; it is involuntary, the sympathetic being the chief motor nerve of the uterus; the uterine contraction is peristaltic, but nearly simultaneous, beginning at the fundus probably;

2. The muscular action of the abdominal walls, which is voluntary in part, partly a reflex involuntary contraction.

The chief expelling force is the contraction of the unstriped muscular fibers of the uterus.

The power of a uterine contraction, together with that of the abdominal muscles is fifty to eighty pounds (Duncan); according to Schatz, seventeen to fifty-five pounds.

II. THE PASSAGES.

The passages comprise: 1. The hard parts, or the bony pelvis. 2. The soft parts, including the uterus, the pelvic floor, and the soft structures which line the bony portion of the parturient tract.

1. OBSTETRIC ANATOMY OF THE BONY PELVIS.

The Pelvis is a strong, bony basin, whose cavity forms the most important part of the birth canal.

The constituent parts of the bony pelvis are the ossa innominata, the sacrum, the coccyx.

The pelvic joints are the symphysis pubis, the sacro-iliac joints, the sacro-coccygeal joint. A slight mobility of the pubic and sacro-iliac joints is usually present in the later months of gestation. Recession of the coccyx occurs during expulsion of the fetal head from the outlet to the extent of one-half to one inch.

The False Pelvis, or Greater Pelvis, is that part of the pelvis above the ilio-pectineal line. It forms with the lower segment of the abdominal wall a funnel-shaped approach to the true pelvis.

The True Pelvis, or Lesser Pelvis is that part of the pelvis below the ilio-pectineal line.

The brim, inlet, superior strait, isthmus, or *margin* of the pelvis is located by the linea ilio-pectinea and the upper border of the sacrum. Its shape is approximately heart-shaped. Sometimes it is oval or more or less rounded.

The obstetric landmarks at the brim are : 1. The promontory of the sacrum, or sacro-vertebral angle. 2. The sacro-iliac joints. 3. The ilio-pectineal eminences, situated at the ilio-pubic joint, on the pubic bone. 4. The symphysis pubis.

The outlet of the pelvis, or *inferior strait,* is lozenge-shaped, and is located by the tip of the sacrum, the sub-pubic arch, and the ischial tuberosities. It is a double triangle whose common base is a line connecting the ischial tuberosities ; the apex of one, the summit of the sub-pubic arch ; the apex of the other, the tip of the coccyx.

The obstetric landmarks at the outlet are : 1. The tip of the coccyx ; 2. The sub-pubic arch, formed by the two descending rami of the pubis ; 3. The ischial tuberosities ; 4. The ischial spines ; 5. The greater and lesser sacro-sciatic ligaments which assist in completing the parturient canal partly formed by the bones.

The greater sacro-sciatic ligaments arise from the posterior inferior spines of the ilium, and from the sides of the sacrum and coccyx, and are inserted into the inner surfaces of the ischial tuberosities. The lesser lie in front of the greater. They arise from the sides of the sacrum and coccyx, and are inserted into the spines of the ischium. The open spaces between the greater and lesser sacro-sciatic notches and the ligaments are respectively the greater and the lesser sacro-sciatic foramina.

The greater sacro-sciatic foramen transmits the pyriformis muscle, and the gluteal, the sciatic and pudic vessels and nerves. The lesser transmits the tendon of the obturator internus muscle and the pudic vessels and nerves.

6. The obturator foramen, bounded by the bodies and the rami of the ischium and pubis. It is closed by the obturator membrane except at the obturator canal.

The cavity of the pelvis is bounded posteriorly, in the main, by the sacrum and the coccyx ; anteriorly, by the pubes, the pubic and the ischial rami ; laterally, by the surfaces of the iliac and the ischial bones.

The posterior wall is smooth, and concave from above

downward. Its depth is four to five inches—five and a half measured on the curve of the sacrum and coccyx. The anterior wall is smooth, and concave from side to side; at the symphysis pubis its depth is one and three-fourths inch. Lateral wall is three and a half inches deep.

Planes of the pelvis. The plane of the brim cuts the ilio-pectineal line and the upper border of the sacrum. In the erect posture the average inclination of the brim to the horizon is about 60°.

The middle plane cuts the middle of the posterior surface of the pubic symphysis and the upper margin of the third piece of the sacrum.

The plane of the outlet cuts the tip of the coccyx, the lower end of the symphysis pubis, and the ischial tuberosities. The inclination of the plane of the outlet to the horizon is 11°, the summit of the sub-pubic arch lying below the level of the tip of the coccyx.

Practically, the plane at which the head escapes from the grasp of the pelvis is a plane cutting the tip of the sacrum and a point a half-inch below the lower end of the symphysis.

Pelvic Diameters and Measurements.

Internal Diameters.

a. At the brim:

True conjugate, from the promontory of the sacrum to the upper end of the symphysis, more exactly to the point at which the symphysis is crossed by the linea ilio-pectinea.

Diagonal conjugate, from the promontory of the sacrum to the summit of the sub-pubic arch.

Transverse Diameter, the greatest transverse diameter.

It cuts the points midway between the sacro-iliac joint and the ilio-pectineal eminence on either side.

Oblique diameters, from the sacro-iliac joints, respectively, to the opposite ilio-pectineal eminence; R. O. from the right, L. O. from the left sacro-iliac joint.

b. At the middle plane:

1. *Antero-posterior diameter*, from the upper margin of the third piece of the sacrum to the middle of the posterior surface of the pubic symphysis.

2. *Transverse diameter*, between points corresponding to the lower margins of the acetabula.

3. *Oblique diameters*, from the centers of each greater sacro-sciatic foramen to the center of the obturator membrane opposite.

c. At the outlet:

Antero-posterior diameter, from the lower end of the pubic symphysis to the tip of the coccyx—practically to the tip of the sacrum.

Transverse diameter, between the tubera ischiorum; the bis-ischial diameter.

Oblique diameters, from the middle of the lower edge of the greater sacro-sciatic ligaments on either side to the point of union between the ischium and pubis on the opposite side.

External Diameters.

External conjugate diameter, or *diameter of Baudelocque*, from the fossa just below the spinous process of the last lumbar vertebra to the most prominent point on the surface overlying the upper portion of the pubic symphysis—a prolongation of the internal conjugate. To locate the spinous process of the last lumbar vertebra draw a line connecting the depressions corresponding to the posterior superior iliac spines. The second spinous process above this line is that of the last lumbar vertebra.

Ilio-spinal or *inter-spinal diameter*, the distance between the anterior superior spines of the ilia measured from the outer borders of the sartorii at their origin.

Ilio-cristal or *inter-cristal diameter* is, in the normal pelvis, the greatest external diameter of the pelvis measured transversely at the crests.

Approximate Measurements of the Static or Dried Pelvis.

Internal Diameters.

Antero-posterior.	Oblique.	Transverse.
Brim, 4 inches.	4½ inches.	5 inches.
Cavity, 4½ inches.	4½ inches.	4½ inches.
Outlet, 5 inches.[1]	4½ inches.	4 inches.

The right oblique diameter at the brim is slightly longer than the left oblique. The average measurements at the brim are more exactly as follows :

Conjugate.	Oblique.	Transverse.
4 inches (10 cm.).	5 inches (12½ cm.).	5¼ inches (13¼ cm.).

The circumference of the brim is about 16 inches (40 cm.), of the outlet 13 inches (33 cm.).

Approximate Measurements of the Dynamic Pelvis.

Internal Diameters.

The internal diameters are all diminished a quarter of an inch by the presence of the soft structures in the dynamic pelvis. The transverse diameter at the brim is still more reduced by the psoas and iliacus muscles, leaving the oblique the longest diameter in the dynamic pelvis.

---

[1] Distance from lower end of symphysis pubis to tip of sacrum, 5 inches ; to tip of coccyx, 3¾ inches ; when coccyx is pushed back, 4½ inches.

External Diameters.

External conjugate	.	.	.	8 inches (20 cm.).
Inter-spinal	.	.	.	10 " (25½ cm.).
Inter-cristal	.	.	.	11 " (28 cm.).

The average external circumference, measured over the symphysis, just below the iliac crests and across the middle of the sacrum, is one yard.

To estimate the internal conjugate from the external, deduct two and three-quarters to five inches, according to the estimated thickness of the bony structures and overlying soft parts.

Difference between the Male and Female Pelvis.

Distinguishing Marks of the Female Pelvis.

As a whole: The false pelvis is wider. The true pelvis is larger in all diameters and of shallower depth. The bones are lighter and more slender. The pelvic inclination is greater. The shape is less triangular.

The *brim:* The sacro-vertebral angle is less prominent. The pubic spines are farther apart.

The *cavity* is less funnel-shaped. The sacrum is shorter and broader, and more strongly curved.

The *outlet:* The width of the sub-pubic arch is greater —75°, the angle in the male being 58°. The symphysis pubis is little more than half the depth of the male.

2. OBSTETRIC ANATOMY OF THE PELVIC SOFT PARTS.

At the *brim* the iliacus and psoas muscles encroach upon the lateral margins of the inlet to the extent of a quarter of an inch, or more, on each side. The external iliac vessels lie on the inner borders of these muscles.

In the *cavity* there are no muscular structures over

the median portion of either the anterior or posterior pelvic walls. On either side of the median portion lie the pyriformis posteriorly and the obturator internus anteriorly and laterally—too thin to affect the pelvic diameters.

The *outlet* of the pelvis is closed by the pelvic floor or diaphragm, which is made up chiefly of muscles and fasciæ.

The Pelvic Floor.

The upper aspect of the pelvic floor is concave; its lower, convex from before backward.

Its upper limit is the peritoneum, except where that structure is lifted off to be reflected over the pelvic viscera and their appendages. Its lower surface is skin.

At its median portion it is obliquely traversed by three muscular slits, the urethra, the vagina, the rectum, all approximately parallel with the pelvic brim, except that the lower end of the rectum turns backward nearly at a right angle with the vagina.

The posterior vaginal wall and the soft structures behind it constitute the sacral segment of the pelvic floor; the anterior wall of the vagina and the soft parts in front of it, the pubic segment of the pelvic floor. (Hart.)

Measurements. Coccyx to anus, in the nullipara, one and three-quarters inch; anus to lower edge of vulvar orifice, in the nullipara, one and a quarter inch—in parous women, one inch—in the primigravida at term, one and a half inch.

Greatest transverse width, on the bis-ischial line, four and a quarter inches.

Perpendicular thickness of the pelvic floor at the anus, two inches.

The average projection of the pelvic floor, below a line drawn from the tip of the coccyx to the lower end of the symphysis, is about one inch.

The length of the sacral segment during labor at the moment of expulsion—coccyx to lower edge of the vulvar orifice—is six to seven inches.

Principal Component Structures.

Levator ani. The most important structure of the pelvic diaphragm is the levator ani muscle. Upon the integrity of this, more than any other structure of the pelvic floor, its strength and support depend. It may be described with approximate accuracy as a hammock-shaped muscle, made up of three fan-shaped fasciculi on either side—the sciatic fan, the coccygeal fan, and the pubic fan. The anatomical relations of these fasciculi are as follows:

1. Sciatic fan. Point at ischial spine; base at the side of the coccyx and fourth and fifth sacral vertebræ (usually described as the coccygeus muscle).

2. Coccygeal fan. Point at tip of the coccyx; base, a line on the fascia (arcus tendineus) extending from the ischial spine to the pubes.

3. Pubic fan. Point at the pubes; base, a line from anus to coccyx in which the fibers interlace with those from the opposite side, behind the rectum. A few fibers go to the recto-vaginal septum just above the anus.[1]

Pelvic fascia. The upper surface of the levator is covered by the internal or superior pelvic fascia. This structure is continuous above with the iliac fascia. It

---

[1] See "Studies of the Levator Ani Muscle." Amer. Journ. of Obstet., September, 1889. Dickinson.

dips down into the lesser pelvis to the arcus tendineus, whence it is reflected over the upper surface of the muscle. Posteriorly this fascia covers the pyriformis, anteriorly the upper portion of the obturator muscle.

The inferior surface of the levator is covered by the perineal fascia. This fascia is divided by the bis-ischial line into two portions, the posterior and the anterior. The posterior, a single layer, invests the corresponding portion of the levator. The anterior is divided into three layers, the deep, the middle, and the superficial perineal fascia. The deep layer covers the lower surface of the levator in front of the bis-ischial line. Between the other two fascial layers lie the transversus perinei, the bulbo-cavernosus and the ischio-cavernosus muscles, which are described as follows:

Transversus perinei. Origin, the ischial tuberosity; insertion, the perineal body.

Bulbo-cavernosus. Origin, the anal sphincter and perineal fascia at one side of it; insertion, by three slips, one into the posterior aspect of the bulb, one into the lower surface of the clitoris, and one into the mucous membrane of the vestibule.

Ischio-cavernosus. Origin, the ischial tuberosity and ischio-pubic ramus; insertion, the crus clitoridis and an aponeurosis overlying the posterior portion of the body of the clitoris.

The *sphincter ani* lies in the plane of the three muscles just described. It is made up of two semicircular bands, each about one-half inch wide, one on either side of the anus. Origin, the tip of the coccyx and adjacent skin; insertion, the tendinous centre of the perineal body.

The *perineal body,* so called, is the mass of elastic and

muscular tissue between the lower end of the vagina and the rectum. Its height is one and a half inch, transverse width, one and a half inch, length of base, antero-posteriorly, one and a quarter inch in the nullipara.

The Parturient Axis.

The *axis of the inlet* is a line perpendicular to the plane of the brim at its central point; its prolongation cuts the umbilicus and the tip of the coccyx. It is continuous with the axis of the uterus at term.

The *axis of the outlet* is the perpendicular to the plane of the outlet at its middle point. If prolonged the axis of the outlet, when the coccyx is pushed back, cuts the lower margin of the first sacral vertebra.

The *axis of the outlet of the soft parts*—the axis of expulsion—looks almost directly forward.

The *axis of the parturient canal* is made up of the axes of the several planes of the parturient canal. Its shape is an irregular parabola.

III. The Passenger.

OBSTETRIC ANATOMY OF THE FETAL HEAD.

For the obstetrician the fetal head presents two general divisions: 1. The cranial vault. 2. The cranial base and face. The former, owing to the semi-cartilaginous character and the mobility of its bones, is plastic, a fact of great importance in facilitating the passage of the head through the pelvis; the latter is firm and unyielding, its bony structures being more highly ossified and more solidly united. Protection is thus afforded during birth to the delicate structures at the base of the brain. It is with the cranial vault that obstetric problems have mainly to deal.

The *bones of the cranial vault* are one occipital, two parietal, two frontal, and two temporal.

The *sutures* are the membranous intervals between two adjacent bones. Of special obstetric interest are the following:

The sagittal or the inter-parietal suture;

The frontal or the inter-frontal suture;

The coronal or the fronto-parietal suture;

The lambdoidal or the occipito-parietal suture.

The *fontanelles* are the membranous spaces between the angles of three or four adjacent bones. The fontanelles of obstetric importance are two: the anterior and the posterior.

The *anterior* or *large fontanelle*, or *bregma*, is situated at the anterior extremity of the sagittal suture. In vaginal examination during labor it is recognized by the following characters:

1. It is kite-shaped or quadrangular, its most acute angle forward. 2. Its average diameter is one inch. 3. Four lines of sutures run into it.

The *posterior fontanelle* lies at the posterior extremity of the sagittal suture. It presents to the examining finger the following distinguishing marks:

1. Its shape is triangular. 2. It is small, usually a mere depression barely perceptible to the finger tip. 3. Three lines of sutures run into it. 4. Behind it is the squamous or triangular portion of the occipital bone, movable upon the basilar portion by a hinge-like joint of fibrous tissue.

Protuberances. The fetal head presents three protuberances which are of interest as obstetric landmarks, viz.: the occipital, the parietal, the frontal protuberance.

The occipital protuberance is one inch or more behind the posterior fontanelle.

The parietal protuberance or boss is the bony eminence at the center of each parietal bone.

The frontal protuberance is the prominence at the center of each frontal bone.

The *vertex* is that portion of the head lying between the fontanelles and extending laterally to the parietal eminences.

The *occiput* is that portion of the head lying behind the posterior fontanelle.

The *sinciput* is that portion of the head lying in front of the bregma.

Average Measurements of the Fetal Head.

The *bi-parietal diameter* is the diameter measured through the parietal eminences; value, three and three-fourths inches ($9\frac{1}{2}$ cm.).

The *fronto-mental diameter* is measured from the summit of the forehead to the center of the lower margin of the chin; value, three and a half inches (9 cm.).

The *occipito-frontal diameter* is measured from the tip of the occipital protuberance to the root of the nose; value, four and a half inches ($11\frac{1}{2}$ cm.).

The *occipito-mental diameter* is measured from the tip of the occipital protuberance to the center of the lower margin of the chin; value, five and a half inches (14 cm.).

The *suboccipito-bregmatic diameter* is measured from

the junction of the nucha and occiput to the center of the bregma; value, three and three-quarters inches (9½ cm.).

The *bi-temporal diameter* is the distance between the lower extremities of the coronal suture; value, three and one-eighth inches (8 cm.).

The *bi-mastoid diameter* is the greatest distance between the mastoid apophyses; value, two and three-quarters inches (7 cm.).

Circumference: The suboccipito-bregmatic circumference is that circumference measured over the junction of the nucha and occiput and over the center of the bregma; value, about thirteen inches (33 cm.) in male—one-half inch greater than in female heads.

Trunk Diameters.

The bis-acromial diameter is four and three-quarters inches. The bi-trochanteric is three and a half inches. The trunk diameters are much more compressible than the cephalic.

Presentation.

Definition. The relation of the long axis of the fetal ovoid to the uterine axis.

Varieties. 1. *Longitudinal.*

        A. Cephalic—including:

          a. Vertex;

          b. Face;

          c. Brow.

        B. Pelvic—including:

          a. Breech;

          b. Feet.

2. *Transverse*, including:

    a. Shoulder;

    b. Arm;

    c. Hand.

The presenting part is that part of the fetal ovoid which offers to the examining finger.

Relative frequency of presentations. In 96 per cent. of all term cases the fetus presents by the cephalic extremity. The presentation is pelvic in 3 per cent. of term births, lateral in 1 per cent. Face and brow cases occur in a little less than $\frac{5}{10}$ per cent. of cephalic births. The preponderance of cephalic presentation is due chiefly to adaptation: the fetus tends to accommodate itself to the shape of the uterus.

Position.

Position is the relation of the presenting part to certain anatomical landmarks at the pelvic brim.

In vertex, face, and breech presentations the long diameter of the presenting part engages in one of the oblique diameters of the pelvis. For each of these presentations, therefore, there are four possible positions. Vertex positions are named according to the particular quadrant of the pelvic brim which the occiput confronts. Thus, when the occiput looks toward the left anterior quadrant, the position is left occipito-anterior; when it looks toward the right anterior quadrant, the position is right occipito-anterior, and so on. Face positions are named in similar manner, according to the direction of the chin; breech positions, according to the direction of the sacrum, and shoulder positions according to the direction of the scapula.

We have therefore the following positions :

Vertex Positions.

    Left occipito-anterior—L. O. A.;

    Right occipito-anterior—R. O. A.;

    Right occipito-posterior—R. O. P.;

    Left occipito-posterior—L. O. P.

Relative frequency : 70, 10, 17, and 3 per cent. respectively.

Face Positions.

    Left mento-anterior—L. M. A.;

    Right mento-anterior—R. M. A.;

    Right mento-posterior—R. M. P.;

    Left mento-posterior—L. M. P.

Breech Positions.

    Left sacro-anterior—L. S. A.;

    Right sacro-anterior—R. S. A.;

    Right sacro-posterior—R. S. P.;

    Left sacro-posterior—L. S. P.

Transverse or Shoulder Positions.

    Left scapula-anterior—L. Sc. A.;

    Left scapula-posterior—L. Sc. P.;

    Right scapula-posterior—R. Sc. P.;

    Right scapula-anterior—R. Sc. A.

Terms right and left refer to the mother.

Posture.

Posture is the relation of the fetal members to one another. The usual posture of the fetus during pregnancy and parturition is one of flexion. Posture as an element in the labor is most important as relates to the head.

CLINICAL AND MECHANICAL PHENOMENA OF NORMAL LABOR.

UNDER normal labor are included, in this manual, only labors in which the mechanical factors are all normal, and which are otherwise uncomplicated—labors, in other words, which possess no element of dystocia. Only vertex births, in anterior position, will be classed as normal.

Stages of Labor.

The first stage, or *stage of dilatation,* ends with the complete canalization of the utero-cervical zone.

The second stage, or *stage of expulsion,* ends with the birth of the child.

The third, or *placental stage,* extends from the birth of the child to the complete evacuation and permanent retraction of the uterus.

Causes of the Onset of Labor.

Of the causes which determine the advent of labor we have little satisfactory knowledge. Probable causes are: The loosening attachment of the ovum; distention of the uterus and consequent reaction of the uterine muscles; development of the contractile powers of the uterus; increasing vigor of the fetal movements; excess of CO_2 in the blood, acting upon the motor centres; growing irritability of the uterus; influence of the menstrual molimen. Separation of the decidua begins with the first active contractions of the uterus. The ovum is thus partially converted into a foreign body. This furnishes a sufficient cause for continued expulsive efforts.

Phenomena of Beginning Labor.

The signs of beginning labor are:
  Lightening;
  Irritability of bladder and rectum;
  Increased flow of vaginal secretions;
  The show, a bloody discharge from the vagina;
  Expulsion of the cervical mucous plug;
  Rhythmical uterine pains.

By lightening is understood the sinking of the uterus, which takes place usually within about ten days before labor is actually established. The uterus settles in the pelvis and the fundus falls forward. The shape of the abdominal tumor is correspondingly altered. The woman becomes smaller about the waist. As the uterus sinks deeper into the pelvis the pressure upon the bladder and rectum is increased, and these viscera evacuate their contents oftener than is the usual habit. Lightening, however, is not observed in all cases.

With the onset of actual labor micturition and defecation become still more frequent, and there is a copious secretion of vaginal and cervical mucus. As labor is established the vaginal discharge may be slightly stained with blood, and sometimes the mucous plug which blocks the cervix during gestation is expelled as a tenacious gelatinous mass.

The most conclusive evidence of beginning labor is the occurrence of rhythmical uterine pains, the uterus hardening with each pain. At first the pains may be little more than a sense of pressure, or pelvic tenesmus, and are felt in the lumbo-sacral region. As labor progresses the pains become more pronounced, extend to the lower abdominal region, and finally radiate down the thighs.

Labor pains. By labor pains are meant the painful uterine contractions of labor. The painful character of the uterine contractions during labor is due to pressure upon the nerve filaments of the uterus and upon the nerve trunks of the pelvic cavity.

The duration of a contraction is thirty to sixty seconds. The intervals between the pains at the beginning of labor are about thirty minutes. They gradually shorten as labor progresses, and may be reduced to a fraction of a minute toward the end of the birth.

The intensity progressively increases, reaching its acme at the moment the head is expelled from the vaginal outlet.

I. Stage of Dilatation.

Dilatation.—Three agencies are concerned in the dilatation of the utero-cervical zone :

1. The traction of the longitudinal muscular fibers of the upper uterine segment ;

2. The hydrostatic pressure of the bag of waters ;

3. The softening of the cervical structures by serous infiltration.

The traction of the upper segment of the uterus tends to draw the lower segment up over the inferior portion of the ovum. Dilatation begins at the os internum. With the first painful contractions the ovum is partially separated from the lower uterine segment, the internal os expands and the membranes are protruded into the cervical zone with each pain, receding in the intervals. As the labor progresses the os internum is permanently obliterated, and the ovum rests against the os externum. From this time the progress of canalization is marked by the expansion of the external os.

The bag of waters is the portion of the membranes which in course of the labor protrudes through the cervix. It plays an important part in the dilatation. Its contained liquor amnii, the *fore-waters*, is partially cut off from that above the head, the *hind-waters*, by the ball-valve action of the head as it is driven into the cervix during a pain. The general intra-uterine pressure, however, is transmitted in some degree to the fore-waters. In accordance with the law of hydrostatic pressure the bag of waters is not only thrust downward, but it exerts a certain amount of expansive action upon the walls of the passive cervical zone. In presentation of the vertex the bag of waters has a watch-glass shape.

The dilatation of the cervix is usually slower and more painful when the membranes rupture prematurely. The fetal head is not so good a dilator as the fluid wedge presented by the bag of waters. It lacks the active dilating power and the equable pressure of the bag of membranes. Still greater is the mechanical disadvantage in malpresentations and malpositions, owing to the greater inequality of pressure upon different parts of the resisting girdle.

The bag of membranes ruptures usually by the time it reaches the pelvic floor, frequently sooner, or only on interference.

Softeniug of the cervix, partially established before the onset of labor, is greatly increased during the first stage. The vessels of the cervix, unsupported by pressure, become engorged during the pains, and a transudation of serum takes place into the cervical tissues.

Retraction ring. During the first stage of labor the upper uterine segment becomes thickened by retraction of the muscular structures into that segment, and the lower segment correspondingly thinned. The line of demarcation between the thickened upper and the thinned lower segment is the *retraction ring*, or, as it is more commonly called, the contraction ring. The situation of the retraction ring is above the brim by the end of the first stage, and it rises higher in proportion to the number and strength of the pains.

Retraction of the pubic segment. The bladder and the whole pubic segment of the pelvic floor begin to be drawn upward during the latter part of the stage of dilatation. The bladder is thus protected from injurious pressure during the birth.

The duration of the stage of dilatation varies from two or three hours to several days. The average duration in primiparæ is fifteen hours, in multiparæ eleven hours.

II. STAGE OF EXPULSION.

The Mechanism of Labor.

The most important phenomena of the second stage of labor are comprised in the series of passive movements which the fetus undergoes in course of its expulsion through the birth canal. This combination of movements is commonly termed the mechanism of labor.

Since the engaging diameters of the head are larger than those of other parts of the fetal mass, the essential mechanical phenomena of the· stage of expulsion are those concerned in the birth of the head. To comprehend them, it must be borne in mind that the fetal head is a body of irregular ovoid shape which in all typical labors tightly fits the passages; and that the parturient tract is a canal whose shape and direction vary from point to point throughout its length. The essential cause of the head movements is accommodation or adaptation of the head to the varying shape and course of the birth canal. These movements are descent, flexion, rotation, extension, restitution, and external rotation.

Descent. In the second stage of labor the uterine contractions are reinforced by the action of the abdominal muscles. Hence the bearing-down character of the pains of this period. Before rupture of the membranes the expellent force is transmitted to the head through the entire uterine contents above it. After rupture, when the fetus has become consolidated under the contractions of the uterus, the propelling force is propagated to the head, mainly, through the trunk.

The head advances with the pains and recedes in the intervals, and this advance and recession continues till the head is well in the grasp of the vulvar ring.

Flexion. A certain degree of flexion exists primarily. It is the normal posture of the fetus in utero. This primary flexion is increased as the descent begins. The head is so hinged upon the neck that the occipito-frontal diameter corresponds to a lever of unequal arms, the frontal being the longer arm. On engagement in the utero-cervical zone, the resistance acting with greater effect on the longer or frontal arm of the lever, the chin dips toward the sternum. Flexion is still further increased when the head encounters the greater resistance of the pelvic canal.

The advantage of flexion is obvious. It substitutes the sub-occipito bregmatic diameter ($3\frac{3}{4}$ inches) for the occipito-frontal ($4\frac{1}{2}$ inches), a gain quite sufficient in most cases to make all the difference between a possible and an impossible delivery. The head is still further accommodated to the passages by the moulding, yet to be described.

Rotation. The longest diameter of the pelvis at the brim is either the oblique or the transverse; at the outlet, the antero-posterior. The head, therefore, as it descends, must rotate through nearly a quarter circle to keep its longest engaging diameter constantly in the longest diameter of the pelvis during its passage through the bony portion of the birth canal.

Rotation of the head is mainly due to the slope of the lateral halves of the pelvic floor downward, forward, and inward. In normal labor the occipital pole first lands upon one lateral half of the pelvic floor, and as it descends it is thrust forward and inward under the pubic arch. Flexion, moulding of the head, and the development of the caput succedaneum, to be described later, favor rotation by increasing the dip of the occipital pole. After the occiput has sunk below the level of the pubic arch its rotation forward is partly due to the fact that this is the direction of least resistance.

Extension. After the occiput has escaped under the pubic arch the nape of the neck rests against the sub-pubic ligament, and, rotating upon the nucha as a pivotal point, the head is born by a movement of extension—the vertex, the forehead and the face successively sweeping over the perineum.   A short pause usually follows the birth of the head.

Restitution. Since the shoulders descend in the opposite oblique diameter to that in which the head engages, rotation of the head during its descent through the pelvis develops a certain amount of torsion of the neck.   After the head is born the neck untwists and the head falls into a position corresponding to that in which it entered the pelvis. This movement is termed restitution, and it is of interest as a means of confirming the diagnosis of position.

External rotation is a still further rotation of the head after birth, due to the spiral movement of the trunk as it traverses the birth canal.

Birth of the trunk. The shoulders and the breech rotate in a similar manner to the head as they descend through the pelvis, but less perfectly.   The rotation takes place in the opposite direction to that of the head, since the shoulders come down in the opposite oblique diameters of the pelvis.   The anterior shoulder is expelled first; or it lodges behind the pubes, and the posterior shoulder first appears at the introitus and escapes over the perineum.   A gush of bloody water usually accompanies the birth of the trunk.

Other Phenomena.

Caput succedaneum. The caput succedaneum is an edematous tumor developed upon the presenting part of the fetus during the second and the latter part of the first stage of labor.   In cephalic presentation it forms upon the part of the head which lies within the girdle of resistance. The vessels of this part, unsupported by pressure during the uterine contractions, become engorged, and serous infiltration of its tissues results. The size of the tumor varies with the severity of the labor. Its location differs with the position in which the head enters the pelvis. In L. O. A. positions it is upon the right, in R. O. A. upon the left posterior-parietal region. In R. O. P. positions it is upon the left anterior, and in L. O. P. upon the right anterior-parietal region. The location, however, may be modified by long-continued pressure in the lower portion of the birth canal after the head has undergone partial rotation.

Moulding of the head. The adaptation of head to pelvis is in part accomplished by moulding. Under the pressure of the pelvic walls the engaging diameters are reduced and the head is elongated in the direction of the passages.

Perineal stage. As the occiput approaches the introitus the sacral segment of the pelvic floor is stretched and thrust downward and forward in front of the advancing head. Its length from coccyx to posterior commissure is increased at the moment of expulsion to five or six inches. The sphincter ani becomes relaxed, the anal orifice gapes widely, and feces are usually expelled from the rectum as the head begins to pass over the perineum.

Pulse and temperature. The maternal pulse-rate is somewhat increased during the pains. The maternal temperature, particularly in hard labor, is usually a degree or more above the normal by the termination of the birth.

The fetal pulse-rate is retarded at the acme of the pains—owing, probably, to increased arterial tension in the fetus.

The duration of the second stage in primiparæ is one to seven hours—average, two hours; in multiparæ, fifteen minutes to two hours—average, one hour.

III. PLACENTAL STAGE.

Events.

    1. Separation of the placenta;
    2. Expulsion of the placenta and clots;
    3. Retraction of the uterus.

The separation of the placenta takes place in the meshy layer of the decidua, and is accomplished partly by contraction of the placental site and partly by the extruding force of the uterine contractions.

The expulsion of the placenta is due to the extruding force of the uterine contractions. The after-birth may present by its amniotic surface or be expelled edgewise. Its expulsion from the vagina is explained by the contractile force of the levator ani muscles. The membranes are gradually peeled up from the uterine walls as the placenta is thrust downward.

The retraction of the uterus consists in a thickening and shortening of its walls, due partly to rearrangement of the muscular fibres, partly to thickening and shortening of the fibres themselves. Normally, retraction of the upper segment becomes complete at the close of labor. Retraction securely ligates the uterine vessels torn across on separation of the placenta. The lower segment remains passive for several hours after the birth of the child.

The duration of the third stage varies from ten minutes to two hours. Its average length is twenty to thirty minutes.

After-pains are painful contractions of the uterus after labor. They occur more commonly in multiparæ than in primiparæ, and are most severe after inertia uteri in the third stage and the consequent retention of blood clots.

The average length of normal labor is, in primiparæ, seventeen hours; in multiparæ, twelve hours.

MANAGEMENT OF LABOR.

PREPARATORY.

THE duties of the obstetrician to his patient in the later months of gestation are scarcely less important than during childbirth and the post-partum period. The enforcement of hygienic rules, attention to the general health, examination of the urine once weekly or oftener during the last two or three months, and instructions with regard to the care of the breasts and nipples, are essential to the proper management of the obstetric case. During this period, too, the physician acquaints himself with the conditions with which he will have to deal in the subsequent care of the patient.

PRELIMINARY EXAMINATION.

(A month before labor.)

Plan.

History.
    General health;
    Character of previous pregnancies, labors, puerperiums; miscarriages, if any.
    Date of last menstruation;
    Important data in the history of the present pregnancy.
Abdominal Examination.
    Pendulous abdomen;
    Hydramnios;

Twins ;

Placenta previa ;

Hydrocephalus ;

Complicating tumors ;

Presentation ;

Position ;

Posture ;

Fetal pulse-rate ;

External measurements of the pelvis, in primiparæ.

Vaginal Examination.

Old injuries—pudendal, vaginal, cervical ;

Placenta previa ;

Obstructing tumors ;

Measurement of the diagonal conjugate and the general capacity of the pelvis.

Method of Abdominal Examination for Presentation and Position.

1. *Preparation.* Place the patient in the dorsal decubitus with the abdomen fully exposed or covered only with a sheet. The hands of the operator should be bathed in warm water to render the sense of touch more acute, and because the contact of cold hands causes reflex contractions of the abdominal and uterine muscles, which interfere with the examination.

2. *Locate the dorsal plane and small parts.* Palpate the entire surface of the abdomen, using light touches or thrusts with the palmar surfaces of the finger tips. Downward pressure on the upper pole in the axis of the uterus steadies the dorsum and brings it nearer to the abdominal wall. Identify the child's back by the length

and breadth of the resisting plane. Distinguish from the lateral plane by the greater width of the dorsal and by the absence of a sulcus between it and the head. Small parts are felt as nodules which glide about under the touch ; their outlines may sometimes be fully traced. Circular rubbing motions with the finger-tips help to identify them.

3. *Examine the lower fetal pole.* With both hands over the lower uterine segment and well apart, finger tips toward the mother's feet, catch the lower fetal pole between the hands.

The head is hard and globular ; it presents a lateral sulcus between it and the trunk ; in primiparæ (not in multiparæ) it lies in the pelvic excavation before labor ; head in the pelvis, the cephalic prominence is greatest on the side of the sinciput.

The breech is, alone, smaller, with all its component elements, larger, than the head ; it lacks the hard and globular character of the head, presents no sulcus, and always lies above the excavation before labor.

The head in one iliac fossa indicates a cross-birth.

4. *Examine the upper fetal pole.* With both hands over the upper segment and well apart, finger tips toward the mother's face, differentiate the fetal poles by the signs already given and by ballottement of the head. The breech lacks the flexible attachment to the trunk which characterizes the head, and has less mobility by reason of this and the greater bulk of its component parts.

5. *Locate the anterior shoulder.* Place the hands over the sides of the head, and, with firm pressure, move them toward the breech. The point at which they

first encounter an obstacle is the anterior shoulder. Identify it, if possible, by its anatomical characters.

Anterior shoulder within one or two inches of the median line indicates an anterior position of the fetal dorsum; anterior shoulder several inches from the median line indicates a posterior position of the dorsum.

6. *Locate the fetal heart.* The point at which it is heard loudest, as a rule, locates, nearly, the position of the lower angle of the left scapula of the fetus, or, at least, the upper portion of the fetal dorsum. Fetal heart in the upper uterine segment indicates a breech, in the lower a cephalic presentation. The heart, however, lies nearly midway between the extremities of the fetal ellipse. In multiparæ, therefore, in whom neither pole sinks into the lesser pelvis before labor, the location of the fetal heart tones is of little or no value for the diagnosis of presentation. Occasionally the focus of auscultation does not lie immediately over the heart. It may be found at some remote point, owing to more intimate contact of the fetus with the uterine walls at that point.

The location of the fetal heart is particularly useful in distinguishing between right and left and between anterior and posterior positions of the dorsum.

External pelvimetry. A convenient pelvimeter for the obstetric bag is Schultze's. Marked asymmetry of the pelvis may be detected by external palpation. Interspinal and inter-cristal diameters both small, indicates general contraction. Inter-spinal equal to or greater than the inter-cristal, indicates flattening. For the external conjugate, seven inches is the average lower limit in normal pelves.

Method of Vaginal Examination and Internal Pelvimetry.

The bladder and rectum should be empty. Antiseptic precautions should be observed, as in examinations during labor. Two fingers should be used. Note depth of the symphysis pubis, width of sub-pubic angle, bis-ischial and antero-posterior diameters, size and shape of the sacrum.

Measure the diagonal conjugate, as follows : Passing the index and second fingers into the vagina, place the outer edge of the tip of the second against the summit of the promontory, the radial edge of the hand against the sub-pubic ligament, and measure the distance between the points of contact.

Find the true conjugate by deducting one-half to three-quarters of an inch, according to the depth and inclination of the symphysis pubis, from the diagonal ; one-half inch when the symphysis measures less, three-fourths inch when the symphysis measures more than one and a half inches.

Estimate the other diameters by palpating the walls of the cavity.

GENERAL PREPARATIONS.

The Lying-in Room should be a large room with sanitary plumbing or none at all, preferably with a southern exposure, and well ventilated. The room, the bedding, and the clothing of the patient must be surgically clean.

Preparation of the Bed.—*Directions for the nurse.* Cover the mattress with a muslin sheet, and that with a rubber sheet large enough to reach across the bed. Spread a clean muslin sheet over the rubber and pin

fast to the mattress.   Spread over that a second rubber
covered with a muslin sheet.   Place two or three fresh
laundered sheets, twice folded, in position to receive and
absorb the discharges.

Labor pad.   Instead of the folded sheets an aseptic
pad of prepared jute or other absorbent material, covered
with cheese-cloth, may be used to receive the discharges.
It should be two and a half to three feet square.   A
large Kelly rubber pad is a convenient substitute for the
absorbent pad.

Obstetric Armamentarium.—The obstetric bag for
ordinary practice should be equipped with obstetric forceps, soft-rubber catheter, hypodermatic syringe, fountain
syringe, uterine douche-tube of glass, needles, needle-
forceps, aseptic sutures, hand-brushes, Sims' speculum,
sponge-holding forceps, volsella, curette, and a yard of
aseptic gauze.

Squibb's chloroform, chloral, Squibb's ergot, morphia
tablets, gr. $\frac{1}{8}$, ext. veratri viridis fl., antiseptic tablets of
the biniodide or bichloride of mercury, or powders as
follows :

    ℞.—Hydrarg. biniodid. ⎫
       Potass. iodid.   ⎬ . . . . āā ℨj.
       Cht. no. viii.   M.
    S.—One to a quart of warm water, as an antiseptic solution.

    ℞.—Hydrarg. bichlorid. . . . . . ℨj.
       Acid. tart. . . . . ℨv.
       Cht. no. viii.   M.
    S.—One to a quart of warm water, as an antiseptic solution.

The nurse should have ready: A dozen clean sheets ;
a dozen towels recently laundered ; a dozen pieces of
cheese-cloth, about eighteen inches square, for wash-
cloths ; two or three pieces of straight unbleached muslin
for binders, a yard and a quarter long by half a yard

wide; two surgically clean rubber sheets, large enough
to reach across the bed—table oilcloth may be substituted
where economy requires; scissors; two dozen shield-
pins of medium size; a bed-pan of earthenware or agate
ironware; two or three clean hand-basins of earthenware
or agate ironware; a slop-jar; one or two new hand-
brushes; plenty of hot and cold water; a yard of strong
linen bobbin, one-sixteenth of an inch wide, for tying
the navel cord; a woollen blanket for wrapping the
child; a child's bath-tub and a bath thermometer; Castile soap; an ounce package of salicylated or borated
cotton for dressing the child's navel; the child's clothing.

ANTISEPSIS.

Antiseptic Agents.

Dry heat at 240° to 260° F. Baking in an oven, ten
minutes for instruments; one hour for dressings.

Boiling or *steaming* for a half-hour. Boiling is best
done in water containing one and a half per cent. of
common washing-soda. This removes greasy matter
and prevents metallic instruments from rusting.

Chemical Disinfectants:

Mercuric iodide solution, 1:2000.

R.—Hydrarg. biniodid. } . . . āā gr. vijss.
     Potass. iodid. }
    Aq. . . . . Oij.   M.

Mercuric chloride (sublimate) solution, 1:2000.

R.—Hydrarg. bichlorid. . . . . gr. vijss.
    Acid. tart. . . . . . . gr. xl.
    Aq. . . . . Oij.   M.

Chlorinated soda solution, 1:10.

R.—Liq. sod. chlorinat. . . . . ʒj.
    Aq. . . . . ʒix.   M.

Creolin solution, 1 : 100.[1]

R.—Creolin. . . . . ℥ijss.
    Aq. . . . . Oij.   M.

Carbolic solution, 1 : 20.[1]

R.—Acid. carbolic. ⎱ . . . . āā ℥jss.
    Glycerin.   ⎰
    Aq. . . . . Oij.   M.

Practical Rules for Disinfection.

Non-metallic utensils may be disinfected with any of the above-mentioned agents; heat is the most effective.

Metallic instruments are best disinfected by dry or moist heat.

Cloths, linen, etc., with any of the agents named except chlorinated soda, which destroys the fabric. Steaming or boiling is best. Dry heat penetrates dressings very slowly.

When the chemical preparations are used, exposure for at least a half-hour is desirable.

The Obstetrician may use a clean apron to cover his clothing and to protect his hands and arms against contact therewith.

Technique of Hand-Cleaning—a. Fürbringer's Method.

1. Clean the nails dry.
2. Scrub the hands and forearms thoroughly with the aid of soap and hot water and a hand-brush for not less than three minutes, giving special attention to the finger tips and the free edges of the nails.
3. Rinse off the soap with clean water.
4. Hold the hands in one of the mercurial solutions (1 : 2000) for several minutes.

---

[1] Approximately.

As an extra precaution the hands may be wet well with ninety-five per cent. alcohol before immersion in the antiseptic solution.

Hand-brushes should be sterilized by steaming for ten minutes, or by boiling in water for the same length of time. The addition of one and a half per cent. of washing-soda to the water helps the removal of fatty matter and makes the cleansing more effective.

b. Permanganate Method.

Steps 1, 2, and 3, as above.

4. Immerse for two minutes in a warm saturated solution of permanganate of potassium in boiled distilled water.

5. Remove the permanganate stain by immersion in a warm saturated solution of oxalic acid, made with boiled water.

6. Wash with sterilized water.

7. Immerse for two minutes in a mercuric iodide or chloride solution, 1:500.

By this method the hands may be rendered sterile to culture tests.

To prevent reinfection of the hands touch nothing that is not aseptic. Dip the hands for a moment in the mercurial solution before each examination.

Lubricants. Use as a lubricant for the hands either the sterilizing solution or a 1:500 solution of mercuric iodide in glycerin. Keeping the hands wet with the biniodized glycerin keeps the skin soft and maintains continuous disinfection.

The nurse should wear wash dresses recently laundered, and should cleanse and disinfect her hands, as the doctor does, before touching the genitals of the lying-in patient.

The patient, at the beginning of labor, should have a bath and an entire change of clothing. Before the doctor's examination the nurse should cleanse the external genitals, the thighs and abdomen of the patient with soap and warm water; then remove the soapy water and wash the parts with one of the mercurial solutions.

In case of foul secretions or gonorrheal discharges the vagina and cervical canal should be cleansed in like manner with soap and water, using gentle friction with the fingers, and following with an antiseptic douche. The object is prophylaxis not only against infection of obstetric wounds of the passages, but also against ophthalmia in the child.

For vaginal use a suitable antiseptic is the chlorinated soda or creolin solution. Mercurial irrigation if used at all should be promptly followed with a plain sterilized water douche to wash out the solution as a precaution against mercurial poisoning.

EXAMINATION OF PATIENT DURING LABOR.

Plan.

1. **Verbal.**
   Precursory signs:
   >Lightening;
   >Irritability of the bladder and rectum.
   Signs of actual labor:
   >Irritability of the bladder and rectum;
   >Expulsion of mucous plug;
   >Bloody discharge—the show;

Rhythmical pains in the lumbo-sacral and lower abdominal regions.

2.   Abdominal.

Pendulous abdomen;

Hydramnios;

Twins;

Placenta previa;

Hydrocephalus;

Complicating tumors;

Presentation;

Position;

Posture;

Fetal pulse-rate;

Bladder, full or empty;

Relative size of head and pelvis.

3. Pelvic.

Condition of the vulva—old injuries, rigidity, edema.

Vagina—well lubricated? old injuries?

Rectum and bladder—full or empty?

Bony pelvis—all diameters, especially diagonal conjugate; shape, inclination.

Cervix—old injuries? dilatable? how much dilated?

Membranes—ruptured or not? watch-glass or glove-finger protrusion?

Presentation;

Position;

Posture;

Caput succedaneum;

Stage of progress.

Diagnosticate vertex presentation by the hard and

globular character of the fetal head, and by the sutures and fontanelles ; position, by locating the sagittal suture and finding which end is forward ; posture, by the relative descent of the fontanelles ; stage of progress, in the first stage, by the degree of expansion of the cervix, in the second, by comparing the situation of the leading pole, occiput, with the landmarks of the birth canal.

Examine all accessible fetal parts, with firm pressure, taking plenty of time, and using one, or, if possible, two fingers. The frequency and strength of the pains and the general condition of the patient, including the pulse and temperature, should be noted.

The statement of prognosis must generally be guarded ; should be made as definite as the facts permit.

MANAGEMENT OF THE STAGE OF DILATATION.

Measures for relief of severe pains are chloral, 3j, in doses of gr. xx every fifteen minutes. Opium, rarely, gr. 1, or equivalent dose of morphia or codeia ; chloroform, by inhalation, rarely, in the latter part of the first stage.

Vaginal examinations should be as infrequent as practicable—every one to four hours. If a careful preliminary examination has been made, a single vaginal examination during the first stage of labor will usually suffice.

Special directions. Active measures for accelerating the first stage are permissible only when necessary to avert danger to mother and child. It is a general rule to *remain with the patient* after the os externum reaches the size of a silver dollar.

The patient should be instructed not to keep the bed, not to bear down with the pains, and to keep the bladder and rectum empty. The lower bowel should in all cases be cleared during the first stage by an enema of warm water. The diet should be prescribed.

The maternal and the fetal pulse-rate should be noted from time to time.

MANAGEMENT OF THE STAGE OF EXPULSION.

Taking the bed. The patient should take the bed at the beginning of the second stage; sooner in case of severe pains or rupture of the membranes.

She should be dressed for the bed with her linen tucked under the arms and pinned, and with a folded sheet fastened above the hips in the manner of a skirt.

Rupture of the membranes. The bag of membranes may be ruptured when it reaches the pelvic floor ; earlier in the second stage in case its persistence retards the labor, causes hemorrhage by separation of the placenta, or otherwise threatens danger to mother or child. Rupture with the finger-nail, or with a stout hair-pin previously flamed, or with clean, sharp-pointed scissors. Pass the instrument with its point resting on the finger as a guard, and puncture during a pain.

Puller. Except in over-rapid labor, the patient may be allowed, during the pains, to pull upon a sheet twisted into a rope and fastened by one end at the foot of the bed.

Obstetric positions. In general, the convenience of the patient should be consulted. For examinations, the dorsal posture is best. During the perineal stage, the preferred position, at least for the primipara, is the lateral.

Vaginal examinations should be as infrequent as will permit a proper knowledge of the character and progress of the labor, one-half to one hour at the most. In strictly normal labor a single examination after rupture of the membranes and engagement of the head is usually sufficient.

Anesthesia. An anesthetic, if rightly managed, may be used with advantage in most cases of labor during at least a part of the second stage. In so-called obstetric anesthesia the aim is merely to dull the pain; not to abolish it. The anesthetic is given, therefore, for short periods and intermittently—with the pains. At the perineal stage, however, it may usually be pushed nearly or quite to the surgical degree. Chloroform is the preferred anesthetic for mere obstetric analgesia. Ether should be substituted when complete anesthesia is required for operations. Ether is equally applicable for partial anesthesia in simple labor, but is less pleasant.

Method. Have the head low and clothing loose, remove false teeth, examine the heart, and protect the skin about the mouth and nose with vaselin or glycerin. For the inhaler use a towel spread over the head and lifted at its middle several inches from the face Ask the patient to breathe rapidly with each pain. Drop on the towel, opposite the face, one or two drops of chloroform at each breath. If ether is used, three or four drops at each respiration will be enough to blunt the pain.

Regulation of the expelling forces. In slow labors the pains may be stimulated by simple means, especially postural measures. The patient should be encouraged to use the abdominal muscles. Over-rapid labor may be moderated by the use of anesthetics and by regulating the abdominal pressure. Anesthetics may accelerate, retard, or arrest expulsion, according to the freedom of dosage. Avoid manipulation of the cervix; it invites sepsis.

Protection of the perineum. The main reliance for preventing injury to the pelvic floor during the birth is the slow and gradual delivery of the head by its smallest circumference. Retard and regulate expulsion by the anesthetic and by direct pressure with the fingers against the occiput. From the time the pelvic floor begins to bulge, the birth of the head should rarely occupy less than a half-hour. Press the head well up in the pubic arch as the forehead is about to escape.

Episiotomy. When extensive laceration is otherwise inevitable, incise the resisting ring at the vaginal outlet bilaterally. Cut during a pain. Pass a narrow, blunt-pointed bistoury flatwise between the head and the resisting girdle. Turn the edge outward and cut horizontally, holding the knife in line with the axis of the patient's body. Avoid the skin. Location of cuts, one-third way from the median line, posteriorly, when the parts are fully stretched. Length of incision, one inch; depth, one-quarter inch. Suture the incisions after labor.

Management of the cord. If wound about the neck, slip it down over the head. If this is impracticable, cut it and deliver the trunk promptly.

Delivery of the trunk. Lift the head well up toward the mother's abdomen and deliver the posterior shoulder first by lifting it over the perineum. Disengage the posterior arm, and then release the posterior shoulder. Extract the trunk slowly or leave to Nature.

Ligation of the cord. In general, wait till notable pulsation ceases at a point near the vulva. By delaying the ligation of the cord till respiration is fully established the child gains from one to three ounces of blood —a matter of importance, especially in premature and in puny and anemic children. Tie firmly with aseptic narrow linen bobbin an inch from the umbilicus. Place a second ligature a few inches farther away. Cut between the ligatures, near the first, with clean scissors. Press the end of the stump with a cheese-cloth dipped in the mercurial solution, to see if it bleeds; if it does, tie again. A thick cord should be firmly pinched before tying, to squeeze out the gelatin of Wharton from the part to be ligated.

MANAGEMENT OF THE PLACENTAL STAGE.

From the moment the head is born keep the hand on the abdomen over the anterior surface of the uterus till evacuation and retraction of the uterus are complete. Use gentle friction, when necessary, to promote normal contractions.

Delivery of the placenta.

Credé's method. After thirty minutes, supplement the uterine contraction, at the acme only, by manual compression, through the abdominal wall, the hand grasping the fundus, thumb in front, fingers behind. Repeat this manipulation with each pain. Don't pull the cord to assist delivery till the placenta lies in the vagina.

Manual extraction. Credé failing, after an hour,

deliver by hand in vagina, seizing the placenta with the fingers passed through the cervix.

Management of the membranes. On expulsion of the placenta twist the membranes into a rope, and continue twisting till they are wholly detached.

Examine the placenta and membranes to make sure they are entirely delivered. Examine the membranes by transmitted light, to see that amnion and chorion are both complete.

LACERATIONS OF THE PASSAGES.

Cervical lacerations should be immediately sutured only in case they give rise to troublesome hemorrhage.

Method of suture. No anesthetic is required. Place the patient in the Sims position on the bed or a table. Expose the cervix with a large Sims' speculum. Draw the cervix well down with a volsella. The traction usually arrests the hemorrhage for the time. Bring the surfaces of the cervical wound together and suture with silk, passing the first suture above the angle of the tear.

Lacerations of the pelvic floor—Frequency. In primiparæ, 15 to 35 per cent. In multiparæ, 10 per cent.

Causes. Narrow pubic arch ;

Small size, relatively, of the vulvo-vaginal orifice ;
Undue rigidity of the pelvic floor structures ;
Advanced age in primiparæ, over thirty years ;
Faulty mechanism ;
Over-rapid delivery ;
Unskilled use of forceps.

Principal varieties. 1. Median laceration. 2. Laceration of the levator ani on one or both sides. Both varieties may coexist. Either may occur subcutaneously.

Degrees of median laceration. First, to the sphincter ani ; second, through the sphincter ani ; third, into the rectum.

Treatment—a. Time for repair. Notable lacerations at the vaginal orifice should, as a rule, be immediately sutured. Union, however, may be obtained by suturing within the first twenty-four hours. The suture may be successfully done any time within a week or more, if the wound surfaces be first vivified by rubbing with a fold of cheese-cloth. Suturing before the delivery of the placenta may save the necessity for renewed anesthesia. It is not advisable in extensive or complicated tears.

b. Suture material. For ordinary sutures, sterilized, paraffined silk ; for buried sutures, sterilized catgut is suitable.

No. 7 silk is a good size for general use. One or two sizes finer may be used for small wounds. Common Corticelli sewing-silk of the dry-goods stores is a good substitute for the usual surgical material. Silk may be sterilized by immersion for an hour in melted paraffin at a temperature between 240° and 260° F. A special thermometer, that may be kept immersed in melted wax, must be used for regulating the temperature. The wax selected for the purpose should be a kind soft enough to become readily plastic at the temperature of the hand. Catgut may be rendered aseptic by Fowler's method of boiling for an hour in absolute alcohol.

c. Needle. A straight or slightly curved Hagedorn or other surgical needle about two inches long is most convenient. A common darning-needle will answer in the absence of a better. It may be held in the fingers or a suitable needle-holder.

d. Anesthesia is usually necessary. Chloroform is generally safe for the purpose, if properly managed.

Slight tears may sometimes be sutured under cocaine anesthesia. The cocaine solution should be sterilized by boiling. Cocaine is most effective when injected at several points into the lips of the wound. Not more than a grain should be used in this manner.

e. Operation. Place the patient in the lithotomy position, with the hips at the edge of the bed. The knees may be held by assistants, or by Dickinson's sheet-sling, as follows : Taking a sheet by diagonally opposite corners, twist it loosely into a rope ; with the patient in the desired position, pass the sheet-sling under both popliteal spaces ; carry one end over her shoulder and across the back of the neck ; pull taut and tie the ends together in front of the chest.

Pack the vagina above the wound with a mass of sterilized gauze, to prevent the flow of blood over the field of operation. Don't forget to remove the packing after completing the suture. Press the wound-surfaces repeatedly with a sponge compress till they are dry. Examine well the character and extent of the injury. Loose tags of tissue that may become necrotic should be trimmed away with scissors.

The suture should re-establish the normal relations of the parts. Place the stitches at half-inch intervals, beginning at the posterior angle of the wound nearest the rectum. In simple median tears enter the needle on the skin-surface close to the edge of the wound, sweep it deeply through one lip, bring it out just short of the bottom of the wound, and pass it symmetrically through the other lip from within outward.

Care is needed to avoid entering the rectum. The course of the suture should be such that when tied the loop shall be nearly circular. Knot the opposite ends of the suture together or hold with catch-forceps. Sutures all placed, tie them tightly enough only to coapt the wound-surfaces, first removing the gauze packing and clearing the wound of clots. Leave the ends an inch in length, to facilitate removal.

In case of torn sphincter ani, bring the ends of the muscle together with two or three special sutures. When the sphincter is completely severed the positions of the ends of the torn muscle are marked by a dimple on either side of the median line, caused by retraction of the muscle. Catch up the ends with a tenaculum and draw them out. Pass each suture through one end, carry it across and pass it through the opposite end.

Tears extending up either vaginal sulcus should be sutured on the vaginal surface to within an inch or less of the skin. The perineal sutures may then be applied in the usual manner. Other tears of the vaginal wall should be closed.

Lacerations entering the rectum may be sutured on three sides—the rectal, the vaginal, and the skin surfaces.

After care. It is not necessary to tie the knees together. The use of the catheter is generally required for a time, at least, after the perineal suture. It should be avoided if possible. Care must be taken to prevent the urine from trickling into the vagina or over the wound-surfaces. The bowels should be opened on the second day, and kept open daily thereafter. Sutures may be removed on the eighth or ninth day.

CARE OF THE PATIENT AT THE CLOSE OF LABOR.

Retraction of the uterus. Watch the uterus for a half-hour or more after the delivery of the placenta, holding the hand flat upon the abdomen over the anterior surface of the uterus. Use gentle friction, if necessary, to promote contraction. One or two doses of ergot are generally required, especially after labors in which chloroform has been given. It is useful as a prophylactic, not only against post-partum hemorrhage, but against puerperal fever, since it tends to prevent the accumulation of blood-clots in the uterus. By diminishing the blood-supply it promotes involution. It may be given by the mouth or injected subcutaneously.

Cleansing. Direct the nurse to bathe, with the mercuric iodide or chloride solution, the external genitals and soiled parts of the patient's body, and to change her linen and bed linen if soiled. Fresh boiled cheese-cloths, not sponges, should be used for bathing.

Vulvar dressing. The external genitals may be covered, after cleansing, with a folded napkin, the *lochial guard*. The dressing must be aseptic, better for the first five days antiseptic. The lochial guards may be made aseptic by boiling for a half-hour; antiseptic by dipping them, after boiling, in the mercuric iodide solution. They should be dried before using.

A good substitute for the napkin as a vulvar dressing may be made of prepared jute or other absorbent material, loosely packed and enveloped in cheese-cloth. The absorbent portion, or pad, should be ten inches long, four inches wide, and two inches thick. A tail about ten inches long should be left at each end of the pad for pinning to the abdominal binder. Burn after using.

Draw sheet. A sheet folded to four thicknesses may be placed under the patient's hips to protect the bed. It should be changed as often as soiled.

The abdominal binder is best made of unbleached muslin, a yard and a quarter long, half a yard in width and without gores. It should reach from the ensiform to a point just below the trochanters; may be moderately tight for the first twelve hours, subsequently looser.

The condition of the mother, including the pulse-rate, temperature, after-pains, and amount of flow, should be noted before leaving.

Instructions to the nurse. Instruct the nurse with reference to the general management of the patient, and especially in the matter of sleep, diet, evacuations of the bladder, nursing the child; also to watch the amount of the lochial flow. Leave a drachm of the fluid extract of ergot to be given in case of hemorrhage, a grain of opium or its equivalent for use if needed for severe after-pains, and a suitable antiseptic to be used in cleansing the genitals. Give directions in regard to the care of the child. For the first few hours the navel should be occasionally examined for possible bleeding.

PHYSIOLOGY OF THE PUERPERAL STATE.

CLINICAL COURSE AND PHENOMENA.

A MORE or less pronounced chill usually follows the birth of the child. It is due chiefly to the lessened heat-production which attends the abrupt cessation of the muscular efforts of labor, and has no pathological significance. The pulse-rate in most cases falls soon after labor below the usual normal standard. For a period of one or two weeks it remains below 60, sometimes as low as 40 to the minute. The low pulse-rate is doubtless to be ascribed mainly to the diminished amount of labor put upon the heart after the birth. The physiological upper limit of temperature in the post-partum period is for the first four or five days $99\frac{1}{2}°$, thereafter 99° F. A rise of one or two degrees may occur, however, on the establishment of the milk secretion, if the breasts are much engorged and painful. Owing to lowered intra-abdominal pressure and other causes the patient is liable to retention of urine during the first one or two days following childbirth. Neglect at this time often leads to overdistention of the bladder. For several days before and after labor sugar may usually be found in the urine. The glycosuria of this period is due to resorption of lactose, and disappears when the balance is established between supply and consumption. Peptonuria is also normal in the puerperal period, peptone being a product of uterine involution.   The bowels, as a rule, act sluggishly.

Condition of the Uterus.   The upper segment is thick and firmly retracted.   The lower segment is thin and flaccid for about twelve hours after childbirth; thereafter it gradually regains its shape and muscular tone.

The cavity.   The inner layer of the decidua remains, to be gradually shed during the lochial flow.   Fragments of the outer, superficial layer are also retained, to be detached and discharged with the lochia.   The placental site is prominent and studded with small blood clots lying in the mouths of the vessels.

Involution is the process by which the hypertrophied structures, especially of the uterus, are restored to the non-gravid condition normal to the parous woman.   It is essentially a process of fatty degeneration, due to the diminished blood supply.   The endometrium is completely renewed.

The *uterus* at the close of labor measures four or five by seven or eight inches externally ; the thickness of its walls is one to one and a half inch ; the depth of the cavity is—

At the close of labor, about	.	.	.	.	.	6 inches.
" tenth day .	.	.	.	.	.	4⅛ "
" second week	.	.	.	.	.	3⅞ "
" third week .	.	.	.	.	.	3½ "
" fourth week	.	.	.	.	.	3⅛ "

After complete involution the thickness, width, and length of the uterus are approximately one, two, and three inches respectively.

In the parous woman the organ remains permanently larger than in the virgin.

The situation of the fundus immediately after labor is nearly midway between the umbilicus and the pubes; a few hours later, just above the umbilicus (and the uterus is usually dextroverted), by the tenth day at the level of the brim. The elevation of the fundus, however, varies with the fulness of the bladder and rectum.

The weight of the uterus at close of labor is approximately thirty-five ounces, end of first week sixteen, end of second week twelve, end of third week eight ounces; after involution, ten to thirteen drachms—one and a half ounce nearly.

The duration of uterine involution is six to eight or even ten weeks.

Involution of the uterus is retarded in non-nursing women, after twin births, premature labor, much hemorrhage, retention of secundines, endometritis, or getting up too soon; also by violent emotional disturbance.

The cervix is soft and shapeless, having an almost jelly-like consistence at the close of labor, and it reaches at least two and three-fourths inches below the retraction ring. After about twelve hours it begins to be gradually re-formed.

The os internum admits two fingers at the end of twenty-four hours; the os externum one finger after seven to fourteen days. Involution goes on *pari passu* with that of the upper segment of the uterus. The lower border is permanently more or less notched in parous women.

The vagina. The vaginal walls are hypertrophied and relaxed after labor. Involution progresses with that of the uterus; they are not completely restored to the nulliparous condition, however.

The uterine ligaments also undergo involution.

After-pains.—The periodical uterine contractions of pregnancy and labor continue for a few hours or days post-partum ; generally they are more or less painful in multiparæ, owing to retention of blood clots, usually not so in primiparæ ; they accomplish and maintain the retraction of the uterus, and are, therefore, conservative. After-pains are usually intensified while the child is nursing.

The lochia are the genital discharges following labor. They are more or less bloody for about four or five days —*lochia rubra*—and contain shreds of decidua and placental tissue; then sero-sanguinolent—*lochia serosa*— for two or three days ; finally, creamy—*lochia alba*—and containing fat-granules, epithelial cells, leucocytes, and cholesterin. Their reaction is alkaline for a week or more, then neutral or acid. Amount, about three and a quarter pounds. Duration, in normal cases, two to four weeks.

MANAGEMENT OF THE PUERPERAL STATE.

Frequency of Visits.—The patient should, as a rule, be seen within twelve hours after labor, and once or twice daily for the first three days ; once daily thereafter until the seventh ; occasional visits should be made during the balance of the month.

Observations at the First Visit.—Note the general condition of the mother, pulse, temperature. Inquire after the amount and character of the lochia. Have the binder loosened and examine the uterus, externally, for size, firmness, tenderness. Observe in the abdominal examination whether the bladder is full or empty. Learn whether the bladder has been evacuated, the quantity of urine passed, and if the patient has had sufficient sleep and proper diet. Look to the child. Ascertain whether it has passed urine and meconium as evidence that the passages are pervious; whether there is any discharge from the eyes, or bleeding from the navel, and what its temperature is, per rectum.

At Subsequent Visits.—Watch the pulse and temperature, the condition of the breasts, nipples, bladder, the lochia, involution of the uterus, and the general condition of the mother. Examine the pelvic contents by the bimanual, once or more during the third or fourth week. Note especially whether the introitus vaginæ is normally closed, the vagina intact, condition of the broad ligaments (exudations or adhesions), cervix lacerated or gaping, the size, shape, position, density, and mobility of the uterus.

The case should not be wholly dismissed until involution is complete and the pelvic organs are restored to their normal non-gravid state. Note the condition of the child at each visit.

Evacuations of the Bladder.—The bladder should be emptied within six hours after labor, and once in six or eight hours subsequently.

Retention of Urine may sometimes be relieved by hot fomentations to the vulva over the meatus, a rectal injection of warm water, suprapubic pressure, and a sitting or half-sitting posture during attempts at micturition.

Evacuations of the Bowels.—The bowels should be opened on the second day, once daily thereafter. Order for this purpose either a simple laxative, an enema of warm water, Oj, or a rectal injection of a saturated solution of Epsom salts, ʒj–ij, or of undiluted glycerin, ʒij.

After-pains may be relieved, if severe, by one or two doses of opium, gr. ½ to j, by phenacetine, gr. v to x, or by chloral hydrate, gr. xx.

Restorative Measures are rest and sleep, as generous a diet as a patient can digest, tonics (iron, quinine and strychnine), and stimulants if indicated.

Antisepsis.—Rigid cleanliness of the patient's person, linen, and bed-linen is imperative.

The lochial guards should be changed every three to six hours during the first three days, and at all times often enough to prevent the slightest putrefactive odor. The external genitals and immediate surroundings and other parts of the body that may be soiled by the discharges should be cleansed with an antiseptic solution on changing the dressing. Vaginal or uterine douches should be used only in case of sepsis or of fetor not controlled by rigid external cleanliness.

The nurse should wear wash-dresses, frequently changed, and should be as careful in the observance of all antiseptic details as the doctor is required to be.

Diet.—The diet should be restricted to liquid or light solid food for the first day, especially if the patient be much exhausted, or has taken an anesthetic. Milk, gruels, beef essence, animal broths, soft-cooked eggs, oysters, custard, oatmeal mush or wheaten grits, dry toast and weak tea or cocoa are suitable. After one or two days, in the absence of exhaustion, fever, indigestion, or loss of appetite, a moderately full diet may usually be allowed.

Tardy Involution of the Uterus.—Measures for promoting involution are, gentle friction for ten minutes, twice daily, hand on the abdomen; galvanism, ten to twenty milliampères, one electrode over upper part of the sacrum, one upon the abdomen over the uterus, sitting ten minutes twice daily; faradism in like manner, ext. ergot, gr. j, t. i. d.; hot vaginal douche, two or three gallons, temp. 115°F., once or twice daily; removal of retained membranes, and curetting in case of hypertrophied decidua.

Use of the Catheter.—Catarrh of the vesical neck after the use of the catheter frequently results from infection carried on the instrument. Catheterism, therefore, should be avoided if possible, and when required should conform to the following rules:

The best instrument is a soft rubber or a glass catheter. The catheter should be boiled for ten minutes immediately before using, and after sterilizing should be handled only with surgically clean hands.

The posture of the patient should be dorsal, with the knees drawn apart. Let the patient or an assistant retract the labia to fully expose the meatus urethræ, and hold them apart until the catheter is passed.

The approaches, including the meatus and surroundings should be carefully cleansed and disinfected.

Pass the catheter, lubricated with sterilized vaselin, about one and a half inch, or until the urine begins to flow. Collect the urine in a cup or small bowl. Repeat the evacuation of the bladder every eight hours. Prevent the entrance of urine into the vagina and its contact with genital wounds. Cleanse the instrument carefully after using.

Regulation of the Lying-in Period.

First week. The patient should keep the bed. Generally after the first few hours she may assume a half-sitting posture, if necessary, for evacuation of the bladder or bowels. *Second week.* She should maintain a recumbent posture on the bed or lounge ; may sit erect in bed during meals and for evacuation of the bladder or bowels. *Third week.* May occupy an easy-chair all or part of the day. *Fourth week.* May have the liberty of the room ; at the end of a month may leave the room.

The duration of the lying-in, however, must obviously vary with the rate of uterine involution and the general progress of convalescence.

Lactation and Nursing.

Colostrum is the thin, somewhat viscid fluid furnished by the mammary glands of the puerpera before the true milk secretion begins. It contains fat globules and has moderate laxative properties.

The true milk secretion is generally established by the third day in primiparæ, the second in multiparæ.

Contra-indications to suckling the infant. Among the conditions which prohibit nursing are recent syphilis, tuberculosis, extreme anemia, epilepsy, poor quality or very deficient quantity of milk, pregnancy.

Signs of deficient lactation are breasts persistently flabby, child not satisfied and showing signs of inanition. Examine the milk. Weigh the child weekly.

Measures for increasing the secretion. Generous diet, milk, tonics, and good hygiene are the best galactagogues. Faradism may help.

Direct the nurse to cleanse the nipples after each nursing with a saturated solution of boric acid in water to which one-eighth part of glycerin has been added. The child's mouth may be cleansed in like manner before nursing. Excessive nursing should not be permitted. The nipple is injured by prolonged maceration.

Gentle massage of the breasts may be permitted in case of simple milk-engorgement; should be prohibited in simple inflammation.

Painful distention of the breasts may be treated with a saline cathartic, abstention from liquids, and the use of a sling or compression binder.

THE CHILD.

Condition at Birth.

The weight of the newborn infant averages from seven to seven and a quarter pounds; males weighing more than females, first less than subsequent births.

The average gain in weight is about one and a half pound per month for the first five months, and a pound per month for the remainder of the first year.

Measurements. See page 41.

The temperature ranges from 98.6° to 99°F., but is easily affected by slight causes.

Circulation. The pulse-rate varies from 120 to 140. The ductus arteriosus, ductus venosus, and the umbilical vein become obliterated in a few days, the foramen ovale generally closing within a few weeks.

Respiration. The lungs are collapsed, and the entire respiratory tract is devoid of air until the first respiratory effort. The air-tract may contain vaginal blood and mucus from premature efforts at inspiration. The first respiratory movement is due partly to air-hunger from arrest of the maternal supply of oxygen, and partly to the reflex muscular contractions caused by contact of cool air with the moist surface of the body. The average rate of respiration is 45 per minute.

The skin of the back and flexor surfaces of the limbs is more or less thickly covered with vernix caseosa, which consists of fatty matter, epidermic cells and sebaceous material. The epidermic layer exfoliates in the first few days, leaving the skin red and irritable.

The bowels. The intestinal contents, *meconium*, consist of intestinal secretions and bile, together with lanugo and epidermic scales derived from swallowed liquor amnii. The meconium is passed off, and the discharges become feculent within the first three or four days.

Genito-urinary organs. The bladder usually contains urine. The urine has a specific gravity of 1005; it contains albumin and sometimes sugar. The testicles have both descended into the scrotum.

Special senses. The sensibility of the skin is feeble at birth, but is fully established during the first one or two days. The newborn infant is sensitive to strong tastes only. It is deaf at birth, since the outer ear is closed and the middle ear is devoid of air. Loud sounds are audible within a few hours, or one or two days. The eye is sensitive to light.

The caput succedaneum disappears within about twenty-four hours, and the head resumes its shape in the course of two or three weeks.

Management of the Newborn Child.

Respiration.—To expand the lungs, provoke deep
inspirations by blowing upon the face, by dashing a few
drops of cold water upon the chest, or by gentle flagellation.

Treatment of Apnœa.

Preparatory measures. Remove the mucus from the
throat with the finger covered with a wet soft linen, or
better by aspiration with a soft rubber catheter. If the
child is pale and collapsed give a rectal injection of
water (℥ij) at a temperature of 105° to 108°F. Guard
against chilling. The normal temperature is best maintained by immersing the child's trunk and lower extremities in water at 98½°F.

Direct insufflation. Place the child upon its back.
Make partial extension of the head by a fold of blanket
under the neck. Cleanse the face and cover with a clean
towel. Prevent inflation of the stomach by pressure of
the hand upon the epigastrium. With the intervention
of the towel, expand the lungs gently by mouth-to-
mouth insufflation. Repeat twenty times per minute as
long as the heart beats.

Sylvester's method. Place the child in the supine
position, with the head well extended by a fold of
blanket under the neck. Draw the arms well above the
head for inspiration. Place them by the sides and
gently compress the thorax for expiration.

Schultze's method. Suspend the child by the shoulders,
face from the operator, feet down, placing a thumb in
front and two fingers over the posterior aspect of each
shoulder, with the index fingers hooked in the axillæ—

inspiration. Relax the pressure of the thumbs to assist inspiration.

Invert the position by swinging the trunk and lower limbs upward and toward the face of the operator, flexing the trunk in the lumbar region—expiration.

In case of feeble infants this method must be used with great caution, if used at all.

Byrd's method. Hold the child supine upon the two hands of the operator, at right angles to the forearms. Tilt the hands by lowering their radial borders—inspiration. Tilt the hands by raising the radial borders—expiration.

Incubation.—Puny infants, especially if premature, may be kept in an incubator for the first month or more, being removed from it only for feeding or bathing. The temperature in the incubator should be kept at about 90°, and gradually lowered to that of the room during the few days preceding the final removal of the child.

Rectal Injection.—Order a rectal injection of a tablespoonful of warm water to be given soon after birth to make sure that the rectum is pervious.

Bathing.—The face should be bathed on birth of the head, and the eyes cleansed and carefully dried as a prophylactic against ophthalmia. The body should be smeared with sweet oil or vaselin to facilitate the subsequent removal of the vernix caseosa.

For the first few months the full bath is best given by immersion. The preferred time is a morning hour midway between feedings. The temperature of the bath should be 98°F. by the bath thermometer; temperature of the room, 75°F. The slightest chilling is injurious.

Duration of the bath ought not to exceed five minutes. A fresh boiled wash-rag of Turkish towelling is cleaner than a sponge. Only a mildly alkaline soap (Castile) should be used, and that sparingly. Particular attention should be given to the scalp. The mouth should be cleansed with water once or twice daily. The full bath may be repeated daily in the summer, about three times weekly in the colder months. Soiled parts of the body must be bathed as often as soiled.

In feeble children the full bath is best postponed for several hours or days. Rubbing daily with sweet oil or vaselin may be substituted. Infant powder is usually unnecessary.

Navel Dressing.—Dress the stump of the navel cord with absorbent cotton impregnated with boric acid, subnitrate of bismuth, or oxide of zinc; turn to the left side, and retain by a loose abdominal binder. Rapid desiccation is the chief reliance for the prevention of putrefactive changes in the stump, and the dressing should be ordered accordingly.

The cord should be dried and re-dressed after each bath; or, after the first bath, rubbing with oil may be substituted for bathing until the cord falls off. This usually happens about the fifth day.

The navel wound must be kept surgically clean. Septic infection of the wound may lead to umbilical phlebitis and pyemia.

Clothing.—The following is a simple and convenient dress for the first half-year or more:

1. The usual napkin of cotton or linen diaper.

2. A flannel undershirt of the softest material, without sleeves, and opening in front.

3. A fine flannel dress with high neck and long sleeves, cut *à la princesse*, opening in front, and about twenty-five inches in length.

4. A muslin slip of same style as the flannel dress.

5. Woollen socks reaching to the knees.

During the night the socks may be removed and the muslin slip replaced with a light flannel night-dress.

All clothing, including the belly-band, should be loose enough to easily admit two or three fingers underneath it. The belly-band is not needed after the navel heals. In all seasons, children of whatever age should wear woollen garments next the skin, and the extremities should be as warmly covered as other parts of the body. No garment ought to be worn until properly laundered.

Nursing.—The child should be put to the breast after the mother has recovered from the shock of labor, generally within eight to twelve hours. Ten to fifteen minutes may be allowed for each nursing.

The usual frequency is once in four hours for the first day or two, then every two hours. Lengthen one interval in the night to four or six hours. Wake the child if necessary on the hour. The intervals should be extended to three hours by the age of three months.

Usually one or more artificial feedings daily will be required after the seventh or eighth month.

Wet-nursing. The nurse should be of mature age, below thirty-five and preferably a multipara. Her child ought to be of the same age as that to be nursed, within one or two months. A menstruating woman is generally unsuitable, a pregnant one always. Sound health is indispensable. Examine especially for phthisis, syphilis, and all contagious diseases. The breasts should be well developed, with prominent veins and well-formed and healthy nipples. Examine the nurse's child to learn whether it has been well nourished.

Weaning. The child should be weaned, as a rule, after cutting eight teeth, except when that period falls in the hot months.

Evacuations of the Bowels and Bladder.—The bowels should move twice, not more than four times daily. Urination occurs every one to four hours. The child's napkins must be changed as often as soiled.

Sleep.—The newborn infant requires from eighteen to twenty hours' sleep out of the twenty-four; should sleep by itself in a crib or cradle.

ARTIFICIAL FEEDING AND INFANT DIETARY.

First Six Months.

For infant feeding the best practicable substitute for human milk is prepared from cow's milk. Either of the following mixtures resembles the natural food of the newborn child.

1. Ordinary Milk Mixture.

Cow's milk—mixed dairy milk— . . .	℥x.
Water, previously boiled	℥v.
Milk sugar (recrystallized and perfectly pure)	ℨvj–gr. xlv.
Common salt	gr. viij.
Lime-water	℥j.

2. Meigs' Mixture.[1]

Cow's milk—mixed dairy milk— . . .	℥ij.
Cream,[2] containing 20 per cent. of fat . .	℥iij.
Water, previously boiled	℥x.
Milk sugar	ℨvj–gr. xlv.
Lime-water	℥j.

[1] As modified by Rotch.
[2] Best, that obtained by the centrifugal machine, since it may be had fresh.

3. Condensed Milk Mixture.

Canned condensed milk	.	.	.	.	.	℥j.
Boiled water	.	.	.	.	.	℥ix.
Cream .	.	.	.	.	.	ʒx.
Salt	.	.	.	.	.	gr. viij.

During the first two or three months any of these mixtures should usually be diluted by adding three to five ounces more water than the formula prescribes.

If mixture 1 or 2 is used the food should be prepared, bottled, and Pasteurized soon after the milk is delivered, in quantity sufficient for the day's consumption.

Mixture 3 may be made fresh just before using. The sweet brands of condensed milk are not objectionable. Food prepared from canned condensed milk by the foregoing formula is not as sweet as human milk.

Method of Pasteurizing. Ten clean bottles[1] are filled to the shoulders, each holding enough for one feeding. The mouths are then lightly plugged with rubber stoppers. Warm the bottles by immersion for a few moments in water at a temperature a little above 100° F. Then stand in a suitable vessel and pour boiling water in the vessel to cover the bottles to the necks. After a half-hour push the stoppers in firmly, remove the bottles from the water, and transfer to the refrigerator for rapid cooling. The same object may be more surely accomplished by keeping the milk for a half-hour at a temperature of 167° F. by the thermometer, and then cooling. Milk thus scalded and promptly chilled remains for at least twenty-four hours sufficiently sterile for practical purposes, and it is saved the injurious chemical

[1] Or as many as the number of daily feedings.

changes that take place in prolonged exposure to temperatures above 167° F. This method has practically replaced full sterilization of milk for infant feeding.

Feeding. The food should be fed at a temperature of 100° F. and directly from the bottle in which it was prepared. Let the child nurse by means of a rubber nipple slipped over the neck of the bottle. The nipple should be boiled for ten minutes before using and the bottles before filling. Both should be carefully cleansed after using.

AMOUNT AND FREQUENCY.

Rules for general guidance.

Age.	Intervals of feeding.[1]	Amount at each feeding.[2]	Number of daily feedings.	Average daily amount.
First day . .	2 hours.	1 drachm.	10	10 drachms.
Second day . . .	2 "	½ ounce.	10	5 ounces.
Third day . . .	2 "	1 "	10	10 "
Second week . .	2 "	1¼ "	10	12½ "
Six weeks . . .	2½ "	2¼ ounces.	8	18 "
Three months .	3 "	4 "	6	24 "
Six months . . .	3 "	6 "	6	36 "

Small and feeble children require to be fed more frequently and in smaller quantities, large and robust children less frequently and in larger quantities than the foregoing table prescribes. The daily allowance required must be determined for the individual case by trial. The stomach capacity, at birth, is approximately

---

[1] Lengthen one interval in the night to 4 or 6 hours.
[2] By measuring glass.

$\frac{1}{100}$ the weight of the child's body. It averages about one ounce at birth and increases about a drachm and a half per week during the first six months. After that time the rate of increase is somewhat smaller. The weekly weight of the child is a good guide to the feeding. A well-nourished child gains at least five ounces weekly during the first five months. If the casein coagulates in hard masses, the proper remedy, as shown by the experiments of Rotch, is further dilution with plain water.

Six to Twelve Months.

Five or six feedings daily, once in three to three and a half hours. Average daily amount, thirty-six to forty-eight ounces. Some farinaceous material, such as barley or oatmeal gruel, may, in most cases, be added to the food with advantage after the sixth or seventh month. The proportion of gruel may be one-eighth the entire mixture.

Undiluted cow's milk mixed with barley or oatmeal gruel and Pasteurized is frequently well borne by healthy children after nine or ten months.

Twelve to Eighteen Months.

Four or five feedings daily of whole milk, sterilized, with barley or oatmeal gruel or bread jelly in the proportions above given. Two or three ounces of raw beef juice, moderately seasoned, may be given daily, either mixed with the milk or separately. It should be prepared at least twice a day and kept on ice. The beef must be fresh.

The simpler kinds of food requiring mastication may be added after the child has sixteen teeth, such as oatmeal and milk, or wheaten grits, well cooked, or stale bread and milk. Scraped beef or soft-boiled eggs may be allowed two or three times weekly.

Eighteen Months to Two Years.

Four or five feedings daily. If the child is hearty a little fine-cut meat may be given with the mid-day meal —such as tender beef, lamb, or chicken.

Milk should be the basis of the feeding until the child has all its teeth, and may constitute a part of it for several years longer. Milk, beef juice, and the farinaceous preparations mentioned afford a sufficient dietary for the entire period of infancy. Proprietary foods for infants are not to be recommended.

DISORDERS OF THE NEWBORN INFANT.

Constipation.

Treatment. Look to the digestion and the feeding. Add cream to the food to raise the proportion of fat to 4, 5 or 6 per cent. This alone generally overcomes the constipation in bottle-fed infants.

Suitable laxatives are the following :

R.—Sod. phosphat. . . . . . . gr. x.
    Sacch. lact. . . . . . . gr. x.
May be given at one dose in a teaspoonful or two of water.

R.—Ext. sennæ fluid. deodorat. (N. F.) . . ʒss.
    Sod. et pot. tart. . . . . . ʒj.
    Glycerin. . . . . . . ʒss.
    Aq. . . . . ad ʒiv.
Dose, a teaspoonful, p. r. n.

Phillips' milk of magnesia, in doses of a teaspoonful, is an eligible laxative for infants.

Useful rectal measures are the injection of clear glycerin, ʒj; sweet oil, ʒiv; or warm water, ʒj. The use of a suppository of soap or cocoa-butter, or a glycerin suppository, generally provokes immediate action of the bowels.

Indigestion.

Symptoms are: flatulence, stools green and curdy.

Treatment. Regulate the nursing or feeding; look to health and habits of the mother. It is sometimes useful to dilute the mother's milk by giving the child a teaspoonful of warm water during the nursing. Pepsin, gr. j, in warm water, ʒj, with each feeding, is of service in certain forms of indigestion.

Colic.

Treat indigestion. For relief of pain useful measures are: chloral, gr. j, in water, ʒj, or in syr. vanil. and water, āā ʒss, once to three times daily, p. r. n.; warm applications to the abdomen; a warm rectal injection, ʒj.

Simple Diarrhea.

Treatment. Look to the feeding and the digestion. A mild laxative as a preliminary may be indicated to remove irritating material; then bismuth subnitrate, gr. iij to v, after each diarrheal movement. This failing, add tinct. opii camph., ♏ iij to vj, to each dose of the bismuth.

Thrush.

The buccal mucous membrane is studded with white patches, due to the presence of a vegetable parasite. The patches are distinguished from milk curds by their firm adhesion.

Treatment. For destruction of the parasite sop the patches every two hours with a saturated solution of boric acid or with a solution of sulphite of sodium, ℥j to ℥j. For the stomatitis which remains after destruction of the fungus, use as a mouth-wash a half-saturated solution of potassic chlorate. Do not allow the child to swallow any of these solutions. The associated digestive disorders are to be treated as in other cases.

Intertrigo.

Use as an infant powder lycopodium and oxide of zinc, equal parts, dusted upon the affected surfaces after cleansing, p. r. n.

Cephalhematoma

is an extravasation of blood between the pericranium and the cranial bones ; it rarely occurs internally. In a few days a hard ridge develops at the margin of the tumor from periosteal inflammation.

Its location is most frequently over one parietal bone, exceptionally at the site of the caput succedaneum.

Prognosis. The prognosis is grave in the internal form if cerebral symptoms develop. The external variety usually terminates in subsidence of the tumor, and recovery in about three months.

Treatment. Ordinarily none. If the tumor grows, shaving the head and firm strapping ; if pus forms, incision.

Icterus.

Benign form. *a.* Due to hematic changes, *i. e.,* destruction of red blood-corpuscles and decomposition of hemoglobin. Conjunctivæ and urine not stained. Very common in the first week. Disappears after six or eight days.

b. Dependent on hepatic conditions peculiar to the newborn child and consequent resorption of bile. Occurs most frequently in feeble infants, and after difficult labor ; conjunctivæ and urine discolored, stools clay-colored in well-marked cases. Duration one to several weeks. Child usually suffers little inconvenience.

Treatment. Usually none required. In persistent cases, attention to the digestion, keeping the bowels open by enemata or possibly by the use of mild laxatives.

Grave form. Rare. May depend on malformation of the bile-ducts, septic infection, interstitial hepatitis from syphilitic or other causes. Distinguished by gradually increasing discoloration, and in septic forms by high temperature, etc. The prognosis is generally fatal.

Treatment must be addressed to the cause.

Ophthalmia.

Cause. Infectious material derived from the genital tract of the mother, generally the gonococcus of Neisser. Begins usually on or before the third day.

The prognosis for the sight is grave in the absence of proper treatment. Twenty-five per cent. of all cases of total blindness are due to this cause.

Treatment. *Prophylactic.* Antiseptic cleansing of the maternal passages during labor in case of leucorrheal secretions ; careful cleansing and drying of the child's eyes immediately on birth of the head. Credé's method : Instillation of one or two drops of a 2 per cent. solution of nitrate of silver into each conjunctival sac directly after birth.

Curative. Applications of cold in the earlier stages, ice-water compresses, in the absence of corneal complications. Removal of the pus every hour or two by bathing and irrigation with a saturated solution of boric acid. After free discharge is established brush the conjunctival surfaces, after cleansing, once or twice daily with a 2 or 4 per cent. solution of nitrate of silver. Continue until the discharge loses its purulent character. Frequent cleansing with the boric acid solution must be continued until all discharge ceases. Anointing the edges of the lids with vaselin promotes drainage by preventing the lids from gumming together. Drill the nurse thoroughly in the method of manipulating.

As a rule, the advice of a competent oculist should be had.

Umbilical Infection.

Most frequently erysipelatous. The cause is uncleanly management of the umbilical wound. It may result merely in a local ulcer, or in umbilical phlebitis and general septicemia. In the latter case the termination is fatal, usually by convulsions. Possible complications are phlegmon of the abdominal walls and peritonitis.

Treatment. In local sepsis, frequent antiseptic cleansing of the wound surface, and dressing with bismuth powder, or iodoform and bismuth. In general septic infection treatment is futile.

Tetanus Neonatorum.

Begins toward the end of the first week. The cause is generally infection of the navel with tetanus bacillus. The symptoms are the same as in surgical tetanus. The termination is almost uniformly fatal within two or three days.

Treatment. Remove, so far as possible, all sources of peripheral irritation. Feed with stomach tube passed through the nostrils, using predigested milk, or this failing feed by the rectum. The drug treatment consists in the use of potass. bromid., gr. iv every four hours, or chloral, gr. j every four hours, guarded with digitalis, p. r. n. ; must be given by the stomach or rectal tube ; sulphonal, gr. iij every two hours, per rectum, has been used with success.

Umbilical Hemorrhage.

May proceed from faulty ligation of the cord, syphilis, icterus, hemophilia.

Treatment. In simple cases re-tie the cord and apply a compress, or transfix the umbilicus with a hare-lip pin and apply a figure-of-eight ligature. In cases due to a dyscrasia treatment is generally futile.

Mastitis.

Inflammation of the breasts is frequently observed in newborn children during the first week. No treatment, as a rule, is required. Should pus form, which very rarely happens, it should be evacuated.

A Bloody Genital Discharge

sometimes occurs in female children in the first few days after birth ; no treatment is needed.

OBSTETRIC SURGERY.

INDUCTION OF PREMATURE LABOR.

Indications.—Certain cases of narrow pelvis, in which the delivery of a living and viable child is thus possible (flattening to between two and three-quarters and three and one-half inches or equivalent contraction); death of the fetus; history of habitual death of the fetus in the last month of gestation, from other causes than syphilis; nephritis of pregnancy, other measures failing; dangerous cases of placenta previa after the period of viability; certain cases of hydramnios, in presence of danger to mother or child.

Methods.

a. Catheterism of the uterus. First step. Hot antiseptic douche against the membranes for twenty minutes. Avoid mercurials.

Second step. Separation of the membranes from the lower segment by means of a uterine sound, or the finger.

Third step. Insertion of an English bougie between the membranes and the uterus.

b. Packing the cervical canal with iodoform gauze. The cervix and vagina should be sterilized before packing. Renew the gauze tampon daily till labor is established.

c. Schrader's method. Alternate applications over the abdomen of hot and of ice-cold water for five minutes each.

d. Glycerin injection. Labor may be established in most cases within two or three hours by injecting about two ounces of glycerin well up toward the fundus, between the membranes and the uterine walls. A convenient injecting apparatus may be made of a No. 10 English catheter and a bulb syringe. The stylet should be cut an inch shorter than the catheter and left in it. The glycerin and the instrument must be sterilized by boiling for an hour, and the passages, especially the cervical canal, should be thoroughly cleansed and disinfected. Rigid precautions must be taken against the injection of air. The injection should be made slowly, to give time for the diffusion of the glycerin.

These manipulations are best practised in the Sims position with the aid of the Sims speculum, the cervix being held forward with a volsella.

To prevent expulsion of the glycerin the patient should be kept for a half-hour or more in the Sims position.

Care of the Child.—Maintain the bodily warmth by aid of artificial heat, best by means of an incubator —Auvard's, Credé's, Rotch's.

An improvised incubator may be made of a wooden box. It should have a removable lid, perforated with five or six half-inch holes at one end, and a false bottom with similar perforations at the opposite end. Heat may be supplied by hot bottles placed in the compartment beneath the false bottom, or by means of a metal water-tank heated by a lamp. A thermometer placed beside the child should register constantly about 90° F. The use of the incubator should be continued for one or two months.

Recourse must be had to gavage, or feeding through a soft stomach-tube, when the child is unable to nurse the breast or bottle or to be fed from a spoon. By incubation and gavage 20 per cent. of children born at the sixth month may be saved. The viability is correspondingly greater in more advanced stages of gestation.

INDUCTION OF ABORTION.

Indications.—Pregnancy nephritis with grave symptoms not yielding to other measures; chronic nephritis;

Most intractable cases of severe vomiting of pregnancy;

Extensive vesicular degeneration of the chorion;

Irreducible retroversion of the gravid uterus;

Absolute contraction of the pelvis, on election of the mother or in conditions unfavorable for celiotomy;

Death of the ovum.

Methods.—Rupture of the membranes, or partial separation of the ovum aseptically, or cervical tamponade with iodoform gauze renewed after twenty-four hours—in urgent cases oftener.

Mechanical dilatation of the cervix to the diameter of three-quarters of an inch and removal of the ovum with curette and forceps is a good method in experienced hands in the first two months. The Ellinger dilator is recommended. The uterus may be emptied in twenty minutes.

These procedures are best practised in the latero-prone position with the aid of a Sims' speculum, the cervix being held with a volsella.

The induction of abortion should be undertaken only with the assent of reputable counsel.

REMOVAL OF AN ABNORMALLY ADHERENT PLACENTA.

Note.—Abnormal adhesions of the placenta may be assumed, as a rule, when the after-birth cannot be delivered entire by ordinary external and internal manual methods after two hours. Mere retention, however, by closure of the retraction ring must not be mistaken for adhesion.

The etiology of adherent placenta is obscure. The cause probably resides in some abnormal condition of the endometrium.

Treatment consists in manual separation and extraction of the placenta with the hand in the uterus. Begin the separation at the portion already detached. Make sure that no fragments remain. Give a hot intrauterine douche of a 2 per cent. solution of creolin, or 1 : 1000 of hydronaphthol. Inject hypodermatically, ext. ergot. fl., ℥ss.

FORCEPS.

The Instrument.—A forceps for general use should be about fifteen inches in length, should have a moderate pelvic curve, an elliptical cranial curve about seven inches in length, and three inches in width externally at the widest part. The space between the tips of the blades when the instrument is closed should be about a half-inch. To permit sterilizing by heat it is best made wholly of metal.

It should be thoroughly cleansed and boiled for half an hour after using ; should always be sterilized, best by

dry or moist heat, before using; should be kept free from
rust and well polished, and the nickel-plating occasionally renewed.

Mechanical Action.—The principal function of the
forceps is traction.

Its use as a lever by means of an oscillating motion
during extraction is a mechanical gain, but is dangerous
to the maternal soft parts. The use of forceps as a lever
is to be condemned.

Compression of the head by forceps is attended with
danger to the child and but slight mechanical advantage
for extraction. In most seizures the compression is
compensated by elongation of another transverse diameter. More may be accomplished by slow delivery, permitting time for moulding of the head by the pressure
of the pelvic walls. The pressure of the blades should, if
possible, be light enough to leave no marks upon the child.

Indications for Forceps.—*Forces at fault.* Cephalic
presentation in which Nature is clearly incompetent to
deliver with safety to mother and child; generally—not
always—when the head has been arrested for a half-
hour after two hours in the second stage.

Emergencies in which immediate delivery is demanded
in the interest of mother or child. This indication may
be present before the head engages. As a rule, version
is to be preferred to forceps before engagement.

Passages at fault. Flattening, not below three and a
half inches in the conjugata vera, or equivalent obstruction; moderate obstruction in the soft parts. As a rule,
the forceps is permissible only after the engagement of
the head. Symphyscotomy is generally better than a
difficult forceps extraction.

Child at fault. Fetal dystocia in certain cases—*e. g.*, arrested occipito-posterior positions, arrested face presentation in favorable positions, moderate hydrocephalus, after-coming head, impacted breech; also, fetal pulse above 160 or below 100. In impacted malpositions of the head, as a rule, forceps should give way to symphyseotomy.

Complicated labor. Certain cases of accidental hemorrhage, prolapsus funis, rupture of the uterus, and of eclampsia, for rapid delivery; or of placenta previa, to bring the head down as a tampon.

Dangers of the Forceps Operation.

a. To the mother. In the low operation, vaginal lacerations and other injuries to the pelvic floor. In the high operation, lacerations of the cervix, the uterus, the vagina, and the muscular structures of the pelvic floor; separation of the pelvic joints; contusions; shock; sepsis.

b. To the child. Injury of the brain from compression, especially rupture of cerebral vessels. Permanent mental and physical infirmities and even death may result from difficult forceps delivery. Temporary paralysis of the facial nerves and deformities of the head frequently occur. The uncleanly, unskilled use of forceps is a dangerous operation for both patients, especially in high applications.

Application of Forceps.

Preparatory measures. Place the patient in the dorsal posture—the American obstetric position, under an anesthetic. Examine the fetal heart before and repeatedly during the operation. Canalization of the cervix should be complete, or nearly so. The membranes should be ruptured and malpositions corrected, if possible. Empty the bladder and rectum. Passages, instrument, and operator's hands and arms must be aseptic. Lubricate the forceps blades with vaselin sterilized by heat, or simply dip them in the antiseptic solution.

Application. Take the left arm of the forceps in the left hand and pass the blade on the left side of the pelvis —between the pains. Hold it at first nearly in a vertical line, and lightly, as you would a pen. Pass two or more fingers of the right hand between the head and the left wall of the passages, the palmar surface inward, pushing the fingers to the base of the skull, if possible. Pass the blade along the palmar surface of the right hand and between the head and the walls of the birth canal, observing both the pelvic and the cranial curves, hugging the head. *Avoid force.* Pass the right blade in a similar manner, the left hand serving as a guide. Adjust the blades in the best possible seizure, as nearly over the transverse diameter of the head as possible. Sink the handles well backward, watching the perineum. If the arms do not lock readily readjust the blades till they do. Never force the locking. Guard against pinching a fold of the vulva in the lock of the instrument. Before using traction re-examine to make sure the blades are correctly applied.

Extraction.—Grasp the handle lightly near the lock, avoiding compression of the head.

Traction should be intermittent—a pull and a pause. The pull should correspond to a pain, if possible, and should last one minute. Reinforce traction by *expressio fœtus* applied by an assistant. Unlock the arms of the instrument in the intervals of traction to relieve pressure on the head.

Guard against slipping. Readjust the blades to a good seizure if they begin to slip. Readjust, if necessary, as the head rotates in the lower part of the passages.

Line of traction. Apply the force in the direction of the birth canal. In order to this, at the brim, grasp the handles with one hand and apply downward pressure with the other resting upon the shanks near the lock. With forceps of moderate pelvic curve, simple traction suffices after the head reaches the pelvic floor.

The direction is practically a straight line parallel with the posterior surface of the symphysis pubis till the head rests on the pelvic floor. The line of traction should then turn almost directly forward. Sweep the handles upward till the anterior edges of the blades hug the ischio-pubic rami as closely as possible, without crushing the intervening soft parts.

The amount of traction force will vary from ten to eighty pounds. *Time* is an important element in a safe forceps extraction. In accordance with a familiar principle of mechanics the resistance of the moving body, and therefore the violence to the maternal soft parts, increases as the square of the rate of motion. At least a half-hour should be consumed in a low forceps delivery, more in a high operation.

Perineal stage. The forceps may or may not be removed during the passage of the head over the perineum.

A half-hour or more should be given to the perineal stage of delivery, except where prompt extraction is required in the interest of the child.

Removal of the forceps. When the forceps is removed

before the birth of the head, the right blade[1] should be withdrawn first, carrying the handle well up over the opposite groin, protecting the soft part with two fingers placed between the ischio-pubic ramus and the anterior edge of the blade; then the left in corresponding manner.

AXIS-TRACTION FORCEPS.

Advantage. Reduces the amount of traction force to a minimum by applying it in the line of descent, and, therefore, to the best mechanical advantage. Permits the normal movements of flexion and rotation as the head descends.

Position of Patient. On a table, the dorsal position; on a low bed, the latero-prone is better.

Application. Adjust the blades to light pressure and hold with the fixation screw.

Traction. Pull at the traction bar. The handles of the forceps serve to indicate the line of traction, which is regulated by keeping the traction rods nearly parallel with the forceps handles. The traction force should never exceed eighty pounds, rarely more than fifty. If preferred, ordinary forceps may be substituted when the head reaches the pelvic floor.

VERSION.

Version is an operation for the complete or partial inversion of the fetal ovoid by manual interference, substituting the cephalic or pelvic pole for a less favorable

---

[1] On the mother's right.

presentation. Cephalic version causes the head to present; podalic, the feet.

Indications.—*a*. For cephalic version. Breech presentation, when all conditions are favorable (external method before labor), shoulder presentation. *b*. For podalic version. Flattening of the pelvis not below three and a half inches c. v., or equivalent contraction of other forms; placenta previa, except in simple cases; prolapsus funis not otherwise manageable; certain face cases before engagement; most complex presentations; shoulder presentation when cephalic version is impracticable; certain emergencies requiring rapid delivery, head not engaged. Contra-indications to version are engagement of the head; high position of the retraction ring; permanent contraction of the uterus, especially in dry labors. Internal version should be undertaken only after the os is fully dilated, or nearly dilated and dilatable. The dead child may generally be delivered by podalic version in higher grades of contraction— three inches c. v.

Dangers of Version.—*To the mother*. In external and bi-polar version the dangers are usually insignificant. Rupture of the uterus is possible in difficult cases. In internal version there is risk of uterine rupture and increased danger of sepsis. Rapid extraction after version increases the danger of laceration and also of shock.

To the child. The dangers to the child are possible fracture of the bones, compression of the spine in internal version, and the usual risks of ordinary breech birth.

A. EXTERNAL VERSION.

External version is practicable, as a rule, only before labor is actually established, and before rupture of the membranes. It is justifiable only when it can be done without violence.

Method. Placing the hands upon the abdomen, one over each fetal pole, push the poles in opposite directions. Operate between the pains. During the pains hold the fetus to prevent reversion to the former presentation. Finally, apply a binder and lateral compresses over the abdomen to prevent recurrence of the malpresentation.

B. BI-POLAR VERSION.

The advantages of the bi-polar over internal version are : a minimum of traumatism and shock ; lessened danger of infection ; the fact that it may be done early in the first stage of labor is an important advantage in placenta previa. The bi-polar should in all cases be preferred to the internal method when practicable.

Method. As a rule, operate with the aid of anesthesia. Empty the bladder and rectum ; place the patient in the dorsal position ; manipulate between the pains ; pass one or two fingers of one hand through the cervix, place the other hand over the opposite fetal pole externally. Push the breech toward the side on which the feet lie. Toss the head out of the excavation into that iliac fossa toward which the occiput points. Toss the trunk in the same direction inch by inch till the knees present. Draw down a knee, or the knees and fect. Complete the delivery as in spontaneous breech cases. A bi-polar manipulation is also applicable for cephalic version.

C. INTERNAL VERSION.

Method. Place the patient in the dorsal position, under an anesthetic, or in the knee-chest position without anesthesia. Protect the clothing of the operator with a sheet or long apron. Pass one hand into the uterus over the abdomen of the child, palmar surface toward the child. Seize a foot, or both feet, and invert the fetal ovoid by traction. If a hand is within reach, snare it and hold it down. The other hand of the operator may be used externally to steady the fundus, or to help the rotation of the child by pushing up the cephalic pole. Relax the hand and desist from manipulation during the pains. To prevent cramping of the hand, operate with the least possible muscular effort.

Except in emergency requiring immediate delivery the birth may be left to the natural forces till the breech is expelled. The completion of the birth is to be managed as in ordinary breech extraction.

OBSTETRIC SURGERY OF THE ABDOMEN.

CESAREAN SECTION; CELIO-HYSTEROTOMY.

Definition.—An operation for extraction of the child by section of the abdominal and uterine walls.

Historical Note.—The operation antedates the Christian era. The earlier operations, however, were post-mortem Cesarean sections done a few minutes after the death of the mother to save the child. The earliest recorded case of Cesarean section upon the living subject was performed in the year 1500.

Capabilities of the Modern Operation.—Timely operations under the Sänger method should save not less than ninety-two to ninety-five per cent. of cases.

Indications.—With a living and viable fetus, the head being of average size, Cesarean section is indicated in flattened pelvis when the conjugate is below two and three-quarters inches (7 cm.), in other forms of contraction in which there is equivalent disproportion between head and pelvis; generally with dead fetus, conjugate below 2½ inches (6.3 cm.), and in cancer of the cervix, when delivery *per vias naturales* is impracticable.

The Preferred Time for Operation is shortly before the expected date of labor. Operation before labor permits better preparation, the patient's condition is better, the uterus retracts as well, and drainage is all-sufficient or can be made so. Certain authorities, however, prefer to wait till labor is established. Before rupture of the membranes there is less traumatism, the child is more certainly alive, and extraction is easier.

Preparatory Measures.—Reinforce the patient's strength before the operation by tonics and hygienic measures. Move the bowels two or three times, a day or two before operating, if practicable. Render the room, table, and all surroundings of the patient aseptic. The temperature of the room should be from 75° to 80° F. The patient should have a total bath and clean linen.

The bladder should be emptied. Make sure there is

no loop of intestines between the uterus and the abdominal walls.

Sterilize the vagina and the cervical canal, the external genitals and their immediate surroundings, by a thorough soap-and-water scrubbing and the subsequent use of the antiseptic solution. Shave the pubes and field of operation. Scrub the surface of the abdomen with soap and water, wash with alcohol, and finally scrub with a 1 : 2000 mercuric iodide or bichloride solution, using sterilized brushes.

Wrap the body and extremities warmly with clean flannels, except the operative field. Cover the clothing about the field of operation with dry cloths or towels, sterilized by steaming for an hour, and finally spread over the patient and immediate surroundings a sheet fresh from the steam-chamber and provided with an opening to expose the field of operation.

Instruments should be sterilized by exposure for fifteen minutes to dry heat at 234° F., or by boiling or steaming for an hour.

The hands and arms of the operator and assistants should be sterilized (permanganate method) and their clothing covered with operating-robes which have been steamed for an hour immediately before using.

Assistants. First assistant should stand on the left of the patient opposite the operator. Another is needed to give the anesthetic. One nurse takes charge of the steam sterilizer and the instruments. Another stands ready to receive the child.

Instruments. Scalpel, straight scissors, thumb forceps, six to twelve catch-forceps—hemostatic forceps, needle-holder and needles, Peaslee's needle, long catch-forceps for holding the sponge-compresses, a large thin-walled rubber tubing (four feet long) as a constrictor for the neck of the uterus, a steam sterilizer for sterilizing cheese-cloths, towels, etc., a dozen medium-sized silk sutures for the deep uterine suture, a dozen fine silk sutures for the superficial uterine suture, a dozen silk sutures of medium size for reuniting the abdominal wound, several dozen gauze compresses to be used instead of sponges.

Summary of the Conditions of Success.—Timely operation; aseptic technique; deep uterine sutures, three to the inch; superficial between the deep sutures; maintenance of the natural temperature of the abdominal viscera; the least possible handling of peritoneal surfaces; operation within thirty to forty-five minutes.

Steps of the Operation.—Median incision of the abdominal wall;

Application of the cervical constrictor;

Median incision of the uterus;

Extraction of the child and placenta;

Closure of the wounds and dressing of the abdominal wound.

Technique of the Operation.—Inject in the thigh, hypodermatically, ext. ergot. fl., ₅ss. Abdominal incision from a little above the navel to a point one inch above the symphysis, uncovering the linea alba. Incise the tendon cautiously, exposing the sub-peritoneal fat. Close bleeding vessels by catch-forceps or ligation before opening the peritoneum. Lift the peritoneum with tissue forceps or catch-forceps, nick it with the scalpel or scissors close to the forceps, and extend the incision to the full length of the abdominal wound, using the finger as a guide.

Pass a loop of the constrictor over the fundus and adjust it around the cervix, tightening it only as required to control hemorrhage. Let an assistant hold the uterus, by means of the constrictor, firmly in the lower angle of the abdominal incision and in central position. Make a short median incision in the uterine wall well above the retraction ring, avoiding the membranes if still unbroken. Lengthen upward with the fingers or scissors, falling short of the fundus. Separate and push aside the edge of the placenta in case of anterior implantation. Plunge the hand through the membranes and extract the child by the head or feet. Clamp the cord at two points with catch-forceps, cut between them, and pass the child to an assistant.

As the uterus slips out of the abdominal cavity hold back the intestines if necessary with hot sterilized towels laid over the upper portion of the incision, or by provisional abdominal sutures. The coverings also help to keep the liquor amnii and blood out of the peritoneal cavity. Keep the uterus wrapped in hot moist cloths. The placenta if not spontaneously separated may be peeled off by grasping it with the hand.

If the membranes were unbroken when the operation was begun, or only recently broken, no antiseptics should be used in the uterine cavity. Avoid irritating the peritoneum by handling, unnecessary sponging, or by contact of chemical antiseptics.

Close the uterine wound with deep silk sutures at intervals of one-third inch, avoiding the decidua. Enter them one-half inch from the incision and pass them obliquely inward. Close the peritoneal coat of the uterus with sutures of fine silk, between the deep sutures. Unite the free surfaces after the manner of Lembert, or let the superficial sutures dip into the muscular coat. Pull down the omentum over the uterus. Remove the constrictor and secure retraction of the uterus, if necessary, by manipulation or faradism.

Close the abdominal wound with silk sutures at intervals of half an inch. Let the assistant draw out the aponeurosis with forceps as the needle is passed, to insure apposition of the cut edges; or, suture separately the peritoneum and the tendon with interrupted catgut sutures, the overlying structures with silk, passed also through the tendon. Give ext. ergot. fl, ʒss, hypodermatically. Dress the abdominal wound with several thicknesses of dry sterilized cheese-cloth. Leave two ounces of iodoform and boric acid—1:8—in the vagina.

After-treatment.—Promote reaction by artificial warmth and by stimulants per rectum, if required. Maintain a rigid cleanliness. Catheterize the bladder every six hours for two or three days. Put the child to the breast as in normal cases. Begin feeding with light liquid food after twenty-four or thirty-six hours. Open the bowels freely with salines given within a few hours after operation. Remove the abdominal sutures by the tenth day.

The patient may usually leave the bed at the end of three weeks. A firm abdominal binder or supporter should be worn for a few weeks after operation.

Post-mortem Cesarean Section.—In case of sudden death of the mother, the child may usually be extracted alive by abdominal section, if delivered within five minutes after the mother's death. It is claimed on good authority that the child may in exceptional cases live *in utero* for several hours after the death of the mother.

PORRO OPERATION: CELIO-HYSTERECTOMY.

Definition.—Extraction of the child by section of the abdomen and uterus, followed by supra-vaginal amputation of the uterus. The tubes and ovaries are also removed.

Note.—This operation was first performed by Edward Porro, of Pavia, Italy, in 1876.

The Results should not differ materially from those of the Cesarean operation.

Indications.—Certain tumors of the corpus uteri, as myomata, etc ; retro-cervical or retro-vaginal myomata; marked puerperal osteomalacia; a septic uterus; hemorrhage after Cesarean section, not otherwise controllable; atresia of the vagina obstructing lochial drainage.

Steps of the Operation.—Preparatory steps and abdominal incision as in the classical Cesarean section. Ligation of the cervix with a finger-thick rubber tube, passing loop over the fundus, first drawing up the ovaries and tubes. Eventration of the uterus. Packing hot cloths about the cervix to keep blood out of the abdomen. Rapid incision of the uterus and removal of the child and placenta. Transfixion of the cervix by two or three knitting-needles or hat-pins passed through the constricting tube and the cervix. Amputation of the uterus three-quarters of an inch above the constrictor. Ligation of the uterine arteries in the stump. Suture of the abdominal wound, stitching the entire circumference of the stump in the lower angle, with the free surfaces of peritoneum in contact. Mummification of the stump with perchloride of iron solution. Abdominal and vaginal dressings as in Cesarean section.

SYMPHYSEOTOMY.

Historical Note.—First done by Jean René Sigault in 1777. After half a century it became obsolete. Revived in Italy by Morisani, of Naples, in 1866. Generally adopted in 1892.

Results.—The maternal mortality of the modern operation is about 12 per cent. Restoration of the symphysis, as a rule, is complete. Complications of the operation are lacerations of the soft parts and hemorrhages.

Indications.—Simple flattening of the pelvis not below two and three-fourth inches in the conjugate, and equivalent disproportion from other causes, irreducible posterior positions of the occiput, impacted mentoposterior face cases, or irreducible brow presentation. The operation is contra-indicated in ankylosis of the sacro-iliac joints. The fetus should be living and viable.

Method of Operating.—With full antisepsis expose the upper end of the symphysis by a vertical incision through the abdominal walls, terminating below at the upper end of the joint. Make a vertical incision between the recti close to the joint and opening into the pre-vesical space. Pass a strong, curved, blunt-pointed bistoury, from above down, hugging the posterior surface of the symphysis, and divide the joint, including the inferior ligament. The left index finger is introduced either into the wound or the vagina as a guide. When, owing to bony ankylosis, or to the sinuous course of the symphysis, division with the knife is impossible, open the joint with a metacarpal saw. The urethra should be held away from the symphysis and to one side by means of a metallic catheter during the division of the joint.

After delivery of child and placenta bring the pubic bones together, holding the urethra and vesical neck backward to avoid pinching the retro-pubic structures between the bones, suture the soft parts, including in the suture the fibrous structures in front of the joint, dress the wound as in celiotomy, and immobilize the pelvic bones with a firm muslin bandage. Two or three straps of rubber adhesive plaster should be tightly applied across the pelvis under the binder above the pubes. Leave in the vagina an ounce or two of iodoform and boric acid, 1 : 8.

After-treatment.—The patient should lie on the back with the limbs outstretched. The urine may need to be drawn off with a catheter for two or three days after operation.

The binder should be changed as often as soiled. The patient may usually leave the bed after three or four weeks, and may dispense with the binder after six weeks.

OPERATIONS FOR REDUCTION OF THE BULK OF THE FETUS. EMBRYOTOMY.

Embryotomy is the general term for all obstetric operations employed to facilitate delivery by lessening the size of the fetus. In a restricted sense it is commonly applied to operations on the trunk only.

Indications. Hydrocephalus too large for safe extraction without perforation and unmanageable by aspiration of the cranial cavity.

Obstructed labor with dead or non-viable fetus, or fetal monstrosity, conjugate above two and a half inches.

CRANIOTOMY.

Definition. An operation for the partial or complete comminution and removal of the cranial bones, to facilitate delivery.

Steps. 1. *Perforation.* Instrument, Smellie's scissors, or Naegele's or other perforator, preferably the trephine. Empty the bladder and rectum. Let an assistant steady the head by grasping it above the brim with the hands placed over the abdomen.

Pass the point of the perforator against the head, perpendicularly to the surface of contact just behind the pubes, using the finger of one hand as a guide and guard. Puncture through a suture or fontanelle, if possible.

Fix the point in the tissues by a screw-like motion, and perforate in similar manner.

Separate the blades in different directions to enlarge the perforation.

The best method of perforating is with the trephine. It removes a button of bone, leaving a permanent opening through which the cranial contents can be readily removed.

The after-coming head may be perforated by making an incision through the skin at the base of the neck posteriorly and passing the ordinary perforator subcutaneously.

Break up the brain throughout with the perforator, and wash out the brain substance with a stream of water from a syringe.

2. *Comminution.* With the craniotomy forceps passed within the scalp, seize the cranial bones, one by one; break them up by rotating the forceps about its long axis, and dislodge them. In moderate obstruction crushing the head with a cephalotribe may be substituted for removal of the bones with the craniotomy forceps.

3. *Extraction.* Remove the bones till the bulk of the head is sufficiently reduced. Guard carefully against laceration of the passages by projecting spicula of bone. Extract the fetus by means of the craniotomy forceps, the crotchet, or, when space permits, with the cephalotribe. In extreme narrowing the cranial base may be delivered edgewise, drawing down the chin. Throughout the operation a strict asepsis must be observed.

CEPHALOTRIPSY.

Cephalotripsy is an operation for crushing the cranial vault. The best instrument for the purpose is a Lusk's cephalotribe.

The application of the instrument does not differ from

that of ordinary obstetric forceps. Care must be used to get a good seizure, while an assistant crowds the head into the excavation. The head should be perforated before applying the cephalotribe. It is then slowly crushed by turning a powerful screw at the handles.

Extract by means of the cephalotribe used as a tractor, guarding against laceration of the passages by projecting spicula of bone. Cephalotripsy is applicable only in moderate contraction.

EVISCERATION.

Evisceration includes all operations for reducing the size of the trunk by removal of its viscera. Perforation of the trunk may be done with the scissors or through the bony coverings of the chest by means of the trephine or common perforator. The viscera are then broken up with the perforator, and removed with stout dressing forceps, or craniotomy forceps, or with the fingers. The bony walls, if necessary, may be removed piecemeal with strong scissors.

DECAPITATION.

Method.—1. *Blunt hook and scissors.* Let an assistant draw the neck firmly down by means of a blunt hook, or a strong tape passed around the neck ; gradually sever the neck with blunt scissors guarded by two fingers of the other hand.

2. *Braun's sharp hook* is a convenient instrument for decapitation.

3. *Écraseur.* Pass a tape around the neck, as follows : Oil it well, knot one end, and pushing the knot over the

neck with the fingers of one hand, catch it on opposite side with the fingers of the other hand; or pass the tape over the neck by means of an English bougie (with stylet) used as a carrier. Pull the chain of a chain écraseur into place by means of the tape. The écraseur is then operated as in ordinary uses of that instrument. A wire écraseur with piano wire or common picture wire may be employed for the purpose, or a chain saw may be used instead of the écraseur.

Extraction.—After decapitation push up the head and deliver the trunk, then extract the head, chin first. Hook two fingers of one hand in the inferior maxilla, and crowd the head through the pelvis by supra-pubic pressure with the other hand. Beware of the danger of rupturing the uterus in these manipulations, and of lacerating the passages by projecting bone fragments.

PATHOLOGY OF LABOR.

ANOMALIES OF THE MECHANISM.

A. ANOMALIES OF THE EXPELLING POWERS.

1. EXCESS; PRECIPITATE LABOR.

The Cause of precipitate labor may be violence of the expelling forces from excessive reflex irritability, or diminished resistance.

The Dangers are for the most part unimportant. Lacerations, shock, and post-partum hemorrhage are the principal dangers to the mother; to the child, asphyxia from the nearly continuous interruption of the utero-placental circulation, and the possible accidents of sudden and unexpected birth.

Treatment.—Moderate the expelling forces by regulating abdominal pressure, and by chloroform. Keep the patient in bed from the onset of the pains.

2. DEFICIENCY : PROLONGED LABOR.

I. PROLONGED FIRST STAGE : TARDY DILATATION.

a. Inertia Uteri: Feeble Pains.

Causes.—Emotional disturbance; full bladder or rectum; imperfect development of the uterine muscle; fibroids; malignant or inflammatory disease of the uterus; anything that lowers the muscular or nervous tone.

Treatment, in the absence of danger to mother and child, should be expectant. Simple inertia uteri calls for no interference so long as the membranes are unruptured and the patient gets sufficient sleep and nourishment. The bladder and rectum should be frequently evacuated and other causes removed, if possible.

Measures for accelerating the first stage, when interference is demanded in the interest of one or both patients, are the following: Keeping the patient on her feet; hot sitz bath; rectal injection of glycerin, ℥ss; alternate use of hot and cold compresses over the abdomen; quininæ sulph., gr. v to x, to rouse the nervous system; faradic current from upper sacral region to posterior vaginal fornix; peeling up the membranes from the lower uterine segment; the passage of an aseptic bougie, or the injection of two ounces of aseptic glycerin between the membranes and the uterine walls after sterilizing the vagina and cervix. Interference within the passages, however, should generally be avoided, if possible.

b. Cramp-like Pains.

Uterine contractions painful but inefficient, more tonic than clonic; failure of the normal changes in the lower segment and cervix which favor dilatation.

Causes.—Neurotic influences; cervical metritis; excessive uterine distention, hydramnios, twins; dry labor and the consequent unequable pressure upon the cervix; malpresentation; too firm adhesion of membranes to the lower uterine segment.

Symptoms.—Excessive pain with no progress in the absence of mechanical obstruction; rigidity of the cervix; after rupture of the membranes, excessive caput succedaneum.

Dangers.—Exhaustion in proportion to the severity of the pains and the loss of sleep and nourishment; in dry labor, pressure effects in case of both mother and child, and septic infection. Atony of the uterus may result. Exhaustion predisposes to prolonged second stage.

Treatment.—Chloral, ʒj, in three doses of gr. xx each, at intervals of fifteen minutes. Opium, gr. j, once or twice repeated at intervals of an hour if required. Chloroform rarely—prolonged chloroform narcosis is dangerous. Rupture of the membranes in hydramnios; peeling them up in undue adhesion.

In dry labor, gradual manual dilatation, after the cervix is obliterated and the os externum thin; the use of Barnes' bags; gentle traction with forceps after dilatation is nearly complete.

Multiple incisions of the cervix when other means fail and immediate delivery is indicated. The latter treatment is rarely permissible, and only when the operator is prepared to suture the incisions if required for hemostasis.

II. PROLONGED SECOND STAGE.

Causes.—Most of the causes which operate in slow first stage: exhaustion; pendulous abdomen; excessive uterine retraction—retraction ring more than half-way from the pubes to the navel; faulty action of the abdominal muscles.

Symptoms.—In neglected cases the temperature and pulse begin to rise, and the passages become hot and dry.

Dangers.—*To the mother.* Exhaustion, pressure-effects, sloughing; after rupture of the membranes, sepsis.

To the child. Chiefly pressure effects.

Treatment.—Exclude obstructive causes by passing the hand into the vagina if necessary. Evacuate the bladder and rectum. Correct uterine obliquity by manual support, posture, or the binder. Summon the aid of the abdominal muscles. Give quininæ sulph., gr. x. Apply hot fomentations to the hypogastrium or the sacral region. Put the woman in the semi-recumbent posture or squatting posture during pains. Ahlfeld's birth stool—two stools so placed as to leave a triangular space between them, opening to the front—may be tried The patient sits over the open space until the head is about to be born.

Use *expressio fœtus*, applied at the upper fetal pole, or to the head only when that pole presents. Push aside intestinal loops, and, with one or both hands laid flat over the abdomen, press downward in the direction of the axis of the inlet.

Ergot should be proscribed as dangerous to the child, and even to the mother. Cramp-like pains should be treated as in the first stage.

Forceps is indicated when longer delay would be dangerous to mother or child; as a rule, when the head has been arrested a half-hour—after two hours in the second stage, especially if the head is low down and there is no recession between the pains.

B. Anomalies of the Passages.

I. Anomalies of the Hard Parts: Deformed Pelvis.

Frequency.—Contraction of some degree is present in ten to fifteen per cent. of all parturients. The higher grades of deformity are rare. Slight contraction is by no means so.

Gravity.—The maternal mortality is three times greater than in normal labor; the fetal much greater than the maternal.

The dangers are those of prolonged labor intensified; operative interference; malpresentation and malposition, which occur more frequently than in normal pelves; prolapsus funis; rupture of the uterus; post-partum hemorrhage.

The minor degrees of deformity are dangerous for the most part to the child only. With early recognition and timely interference they present little difficulty.

General Character of the Anomaly.—Rarely the abnormity consists in faulty inclination only. In the majority of narrow pelves the contraction is at the brim, and is most frequently an antero-posterior flattening. Obstruction may arise from fractures, exostoses or other tumors.

DESCRIPTION OF FORMS.

Flattened Pelvis.

a. Non-Rhachitic. The commonest variety of reflex contraction is a simple antero-posterior flattening. Relation between inter-cristal and inter-spinal diameters as in the normal pelvis. A false promontory is sometimes present at the second sacral vertebra. The true conjugate rarely falls below three inches.

b. Rhachitic. Inter-spinal and inter-cristal diameters nearly equal. Bis-ischial diameters increased. Pubic arch widened, vertical diameter of pelvis diminished, sacrum flat or convex from side to side, promontory jutting strongly forward.

Flattened and Generally Contracted Pelvis.

a. Non-Rhachitic. Has the characters of a justo-minor pelvis, together with a sinking of the sacrum between the ilia. Promontory low, symphysis short.

b. Rhachitic. Shape of the brim triangular. Subject of small stature.

Justo-Minor Pelvis: Pelvis æquabiliter Justo-Minor.

A generally contracted pelvis. Diameters not, in all cases, uniformly contracted. In occasional cases, the narrowing is confined chiefly to the outlet. Its size bears no relation, necessarily, to the size of the body. Much less common than the foregoing forms of contracted pelvis.

Funnel-shaped Pelvis.

Male pelvis. Pelvis narrowed at the outlet; tubera ischiorum are approximated; antero-posterior diameter at the outlet may be shortened. Sub-pubic angle narrow. Sacrum long and slightly curved.

Kyphotic Pelvis.

Upper end of the sacrum tilted backward. Pelvic inclination diminished.

Transverse diameter increased in the false pelvis, diminished in the true pelvis; pelvis somewhat funnel-shaped ; sacrum narrowed, its longitudinal curvature diminished, transverse curvature increased ; pubic arch narrow, symphysis prominent; the cause is kyphosis in the lumbo-sacral region.

Naegele Oblique Pelvis: Ankylosed Obliquely Contracted Pelvis.

Ankylosis of one sacro-iliac joint and narrowness of the corresponding half of the pelvis. Shape of the brim an oblique oval, symphysis not opposite the promontory. Walls of the pelvic cavity converge below, sacrum asymmetrical, pubic arch narrow.

Ordinary Oblique-ovate Pelvis.

Shape similar to that of Naegele, but deformity due to coxitis ; contraction on the healthy side.

Robert's Pelvis.

Ankylosis of both sacro-iliac joints, and marked transverse contraction from imperfect development of the sacrum ; exceedingly rare.

Spondylolisthetic Pelvis.

This deformity is due to a gliding forward of the last lumbar on the first sacral vertebra. The inferior surface of the former finally rests upon the anterior surface of the latter and becomes firmly united to it. The result is extreme shortening of the conjugate. It is a very rare deformity.

Osteomalacic Pelvis.

In osteomalacia the softening bones yield in the direction of the existing pressures. The osteomalacic pelvis is, therefore, sometimes termed the compressed pelvis.

Narrowing of the Pelvis from Bony Tumors.

This form of obstruction comprises simple exostosis, callus or displacement of the bones due to fracture, and malignant bony growths.

DIAGNOSTIC SIGNS OF ·PELVIC DEFORMITY.·

Clinical data. Evidence of rhachitis, such as tardy dentition, sweats, pigeon-breast, curvature of tibiæ or spine, large joints, very low stature ; deformities in near relatives ; character of previous labors ; pendulous abdomen ; presenting pole persistently above excavation during labor.

Pelvimetry. This is often the only means of diag- · nosis.   (See page 123.)

MECHANISM OF LABOR IN FLAT PELVIS.

The head passes the brim with its long (occipito-frontal) diameter in the transverse diameter of the pelvis, and with the sagittal suture level or nearly so.

MANAGEMENT OF LABOR IN FLAT PELVIS.

Conjugate three and a half inches or more.   The spontaneous delivery of a living child is possible in the majority of cases.   Preserve the membranes—by colpeurynter if required.   Correct malpositions.   Keep bladder and rectum empty.   Regulate the pains.

When Nature fails, deliver by

1. Forceps—Tarnier—provided the head is engaged, child living `and viable.   Forceps is here much more dangerous to mother and child than in the normal pelvis.

2. Version (podalic) when head is not engaged, child living and viable, and other conditions favorable.

3. Craniotomy in case the child is dead and extraction by forceps or version would be difficult.

Conjugate two and three-quarters to three and a half inches. Symphyseotomy if the fetus is living and viable; version or craniotomy in case of dead or non-viable fetus.

Conjugate two and three-fourths inches or less: absolute contraction. Cesarean section, or in certain cases, the Porro operation. (See Porro Operation.) Where the deformity is discovered in time, the claims of early artificial abortion should be considered.

The choice of operation, however, in narrow pelvis, must be based on the relative rather than the actual size of the pelvis; in other words, upon the degree of disproportion between the head and pelvis. The size of the head should be carefully estimated by palpation through the abdominal walls, and by ascertaining the depth to which it has sunk or can be made to sink into the excavation, before deciding the choice of procedure for delivery. During labor, after sufficient dilatation, the half-hand may be passed into the uterus if necessary for determining the relative size of head and pelvis. In moderate contraction gentle traction with forceps may be tried for the same purpose.

MANAGEMENT OF LABOR IN OTHER PELVIC DEFORMITIES.

The choice of operation must depend upon a careful estimate of the character and degree of obstruction. At term Nature is competent in a small proportion of cases. Alternatives in slight contraction are version before, forceps after the head has engaged. The possibilities of a living birth by premature labor may be considered when the conditions are known in time.

Symphyseotomy may be done with a conjugate above three inches and but moderate contraction in other diameters. Generally craniotomy best serves the interests of the mother in case of dead or non-viable fetus. In the higher grades of obstruction Cesarean section or the Porro operation is indicated.

II. ANOMALIES OF THE SOFT PARTS.

Vulvar Atresia.—Atresia may proceed from inflammatory adhesions of the labia majora, edema vulvæ, thrombus, cancer, simple rigidity, rigid hymen.

Treatment. Thrombus rarely requires incision, evacuation of the clots and styptic packing. Nature or forceps is generally competent. A rigid hymen should be incised at one or several points. Other forms of rigidity are generally manageable by forceps with perhaps the aid of episiotomy.

Vaginal Atresia.—May be congenital or acquired, may be annular or may involve the entire length of the canal. In annular atresia multiple incisions and delivery by forceps will usually be required; in complete atresia the Cesarean or Porro operation.

Cystocele may be replaced after catheterizing. The use of the catheter being impossible, aspirate.

Rectocele is replaceable in the latero-prone or genupectoral position.

Rigidity of the Cervix may arise from atrophic changes in aged primiparæ, hypertrophy of the portio vaginalis, or cicatrices.   The dilatation should be left to Nature, except in the presence of danger to mother or child.   Artificial measures, when deemed advisable, are Barnes' bags, manual dilatation, rarely deep cervical incisions.

Cancer of the Cervix.—Premature labor, cervical incisions through the healthy tissues by thermo-cautery, or extraction by forceps, are the principal reliance for delivery in the later months.   The passages should be repeatedly douched with an antiseptic solution during and after the labor.   Mercurials, however, must not be used.   In extreme cases Cesarean section may be required.   It is best done before labor.   The entire uterus may then be removed if the disease has not invaded the parametrium and the condition of the mother permits. When the disease is discovered in the early months, hysterectomy should be considered in accordance with the usual rules in non-gravid cases.

Occlusion of the Os Externum.—The treatment consists in reopening the os by incision from behind forward.

Tumors.—*Treatment.* *a.* Vesical calculi. Replacement, or, this being impossible, remove by vaginal lithotomy.

b. Vaginal tumors. Removal if practicable, otherwise Cesarean section or the Porro operation.

c. Uterine tumors. Pedunculated tumors may sometimes be pushed above the head with the aid of the genu-pectoral position, or removed with écraseur or scissors. The Cesarean or Porro operation may be required.

d. Ovarian cysts. Ovariotomy during pregnancy; during labor reposition or aspiration through the vaginal fornix, or Cesarean section at the beginning of labor or shortly before, with ovariotomy when not otherwise manageable.

C. Anomalies of the Passenger.

OCCIPITO-POSTERIOR POSITION.

In most cases in which the head offers primarily in occipito-posterior position, the occiput rotates to the front, either above the brim, in the cavity, or at the vaginal outlet. In a certain proportion of cases the sinciput rotates to the pubes and the head is born with the face to the pubic arch. In this position the long diameter of the head does not conform fully to the axis of the pelvis and the labor is impeded. Not infrequently in neglected posterior positions of the occiput, the head becomes arrested by impaction in the pelvic brim.

The Causes of anterior rotation of the sinciput are : imperfect flexion, occiput and sinciput striking the pelvic floor at about the same time; diminished resistance of the pelvic floor, and consequent failure to shunt the occiput forward; certain deformities of the pelvis, especially general contraction, oblique deformity, and kyphotic pelvis, disturbing the normal mechanism.

The Dangers in persistent occipito-posterior position are: To the mother, exhaustion; laceration of the perineum. To the child, those of prolonged second stage—mortality 15 per cent. In relatively large pelves the malposition is practically unimportant.

Diagnosis.

Abdominal Signs:

No dorsal plane;

Small parts in middle section of the abdomen;

Cephalic prominence, marked ;

Heart heard over lateral aspect of abdomen well toward the back.

Anterior shoulder remote from the median line.

Vaginal Signs. Large fontanelle easily accesible indicates either an occipito-posterior position or an imperfectly flexed anterior position. Distinguish by relative situation of fontanelles.

Treatment.—*a. Above the Brim.* Before rupture of the membranes the patient should lie in a latero-prone position, on the side to which the occiput points ; anterior rotation of the dorsum is thus often possible. A genupectoral position is a still more effectual means for promoting the normal mechanism. This failing, after sufficient dilatation rotate by combined internal and external manipulation. One hand should be placed upon the mother's abdomen over the child's trunk ; the fingers of the other are passed into the uterus to the posterior shoulder of the fetus. In this manner the child's dorsum as well as the occiput may be brought to the front. When the head alone is rotated it is seldom possible to prevent its return to the posterior position.

b. In the Cavity. Anterior rotation of the occiput may be promoted by keeping the patient upon the side which the occiput confronts ; by the use of upward pressure against the sinciput during the pains to help flexion ; sometimes by manually assisting rotation. If the head becomes impacted, axis-traction forceps should be tried. Head immovably fixed, symphyseotomy may be considered.

c. *At the vaginal outlet* the occiput may almost invariably be rotated into anterior position by backward pressure with the fingers placed against the anterior temple, combined if necessary with forward pressure upon the occiput. Rarely only must the head be delivered in occipito-posterior position.

FACE PRESENTATION.

The frequency of face presentation is 1 : 250.

Causes.—The extension of the head is developed during the labor. The causes are: narrow pelvis, brim narrowed by prolapsed extremity, dolicho-cephalus (excessive length of the occipital portion of the head), large child, enlargement of the neck or thorax, excessive obliquity of the uterus, pendulous abdomen, mobility of the fetus from small size or from excess of liquor amnii, impaction of the occiput in occipito-posterior positions. The preponderance of left mento-anterior positions is explained by the right obliquity of the uterus.

Mechanism.—The head descends with its occipito-mental diameter in relation with the axis of the birth canal, but with that diameter inverted, mental pole first. The values of the engaging diameters of the head are the same as in vertex presentation, three by three and a half inches or a little more. The difficulty of face births is due to the fact that the thickness of the neck and a portion of the chest is added to the long diameter of the face as the face descends, making a total diameter of six and a half inches.

Positions:

    Left mento-anterior—L. M. A.
    Right mento-anterior—R. M. A.
    Right mento-posterior—R. M. P.
    Left mento-posterior—L. M. P.

Mechanism of Mento-anterior Positions.

Head Movements:

1. *Extension.* Corresponds to flexion in vertex births.

2. *Rotation.* Unlocks the difficulty of face birth. Failure is more serious than in vertex presentation. The mechanism is the same as in vertex births (*mutatis mutandis*).

3. *Flexion.* Corresponds to extension in vertex deliveries. The head rests by the lower surface of the inferior maxilla upon the margins of the ischio-pubic rami as pivotal points, and is expelled by a movement of flexion, the face, the forehead, the vertex, and the occiput successively sweeping over the perineum.

4. *Restitution.*

5. *External Rotation.*

Mechanism of Mento-posterior Positions.—In a typical case the birth of a persistent mento-posterior position is impossible, since it would require a diameter of six and a half inches to pass through the pelvis. Spontaneous rotation of the chin to the front is generally possible.

Dangers of Face Presentation.—To the mother, exhaustion and pressure effects. To the child, cerebral congestion from pressure on the veins of the neck. Rotation failing, nearly all die.

Prognosis.—Mortality, 6 per cent. of the mothers,

13 per cent. of the children. The face of the child is usually much disfigured.

Diagnosis.

Abdominal Signs:

Hour-glass shape of the uterus;

Cephalic tumor very round and filling only one side of the pelvis;

Cephalic prominence in relation with the back and generally on the same side of the median line with the breech;

Sulcus at the junction of the head and back;

Heart and small parts on the same side;

Inferior maxilla accessible to palpation.

Vaginal Signs:

Orbital ridges;

Nasal bones;

Malar bones;

Alveolar processes;

Chin.

Treatment.—Nature is competent in a large proportion of mento-anterior presentations, and in most mento-posterior presentations that rotate. In cases seen before engagement of the face, however, as a rule, the malpresentation should be corrected. In order to this, proceed as follows:

Preserve the membranes if possible until full dilatation. Then while the head is still movable at the brim convert the presentation into a vertex by one of the following methods.

1. Schatz: Thrusting the breech forward[1] with one

_____
[1] Toward the child's feet.

hand, the chest backward and upward with the other, by external manipulation, and finally crowding the breech downward. Applicable only before rupture of the membranes.

2. Baudelocque: Pushing up with the fingers first against the chin, then the fossæ caninæ, then the brow, by internal manipulation.

3. Both combined, with the help of an assistant.

Since the conversion of a mento-anterior face case into a vertex presentation results in an occipito-posterior position, the operation should be completed by rotating the fetus into an anterior position.

Prolapsus funis calls for version.

In the cavity it may be necessary to promote extension and assist rotation by drawing the chin forward during a pain, especially in mento-posterior cases.

Impaction or delay in mento-anterior positions is usually manageable by forceps.

In mento-posterior positions, face immovably fixed, fetus living, deliver by symphyseotomy; fetus dead, by craniotomy.

BROW PRESENTATION.

Brow presentation is a semi-extension of the head. Persistent semi-extension is rare; partial extension is generally converted spontaneously into a vertex or face.

The frequency of brow presentation is about 1:1800.

The causes are substantially as in face presentation.

The positions are the same as in face presentation.

Prognosis.—Delivery in persistent brow cases is possible only with a relatively large pelvis. Maternal mortality, 1:10; fetal, 1:3.

Diagnosis.—*Abdominal Signs.* Those of face presentation imperfectly developed.

Vaginal Signs. Orbital ridges on one side, bregma on the other side of the presenting part.

Treatment.

a. Before engagement, conversion into vertex by seizing the head, pushing it up and hooking down the occiput, hand in the vagina, under anesthesia. The fundus should be supported by firm pressure with the external hand. External pressure over the occiput helps the manœuvre.

b. Conversion into face by traction on the upper maxilla with the fingers; not advisable in mento-posterior positions.

c. Version for rapid delivery if indicated in interest of mother or child, and the head is not engaged or the uterus firmly contracted.

d. Symphyseotomy in impacted and irreducible brow cases if child is alive and viable Craniotomy, in dead fetus.

BREECH PRESENTATION.

Varieties.—Breech, knee, footling.

Frequency.—1:60, exclusive of premature births.

Causes.—Narrow pelvis; uterine tumors; placenta previa; hydrocephalus; multiple fetus; conditions favoring mobility of the fetus, such as, multiparity, prematurity, lax uterine walls, hydramnios, shape of uterus possibly, small fetus.

Mechanism.—Usually the bis-iliac diameter engages in one oblique diameter of the pelvis.

Positions: Left sacro-anterior—L. S. A.

Right sacro-anterior—R. S. A.

Right sacro-posterior—R. S. P.

Left sacro-posterior—L. S. P.

Rotation in breech is not so marked as in head presentations. As the breech descends the anterior hip first lands upon the pelvic floor and is born first. The shoulders rotate more or less perfectly. The head rotates as in vertex births. In dorso-posterior positions the occiput almost invariably comes eventually to the front. Spontaneous expulsion of the after-coming head is exceptional.

In persistent dorso-posterior positions, as in anterior positions, the head is generally delivered mental pole first. . The chin catching upon the pelvic brim, delivery may be accomplished occiput first.

Prognosis.—To the *mother:* First stage of labor may be more tedious. Second stage is frequently more rapid. In artificial delivery, laceration of the cervix is more frequent than in vertex births; laceration of the perineum is the rule, especially in first labors. The risks to life are not increased.

To the *child:* The mortality, without interference, is one in three or four; with skilled management but little greater than in vertex births.

Cause of the fetal mortality is apnea from impeded blood supply due to retraction of the uterus after expulsion of the trunk, and from compression of the funis after the head engages. The danger is increased in dry labor.

Indications of danger to the child after delivery of the breech are: funic pulse irregular and feeble; occasional gasping respiratory movements; convulsive movements.

Diagnosis.

Abdominal signs:

Fundal pole hard, globular, ballots, sulcus between it and trunk;

Lower pole irregular in shape, not so hard, in primiparæ above excavation before labor.

Vaginal signs:

Glove-finger protrusion of the membranes;

Absence of the hard globular head;

Absence of fontanelles and sutures;

Tuberosity of the ischium;

Tip of the coccyx, anus, genitals, on a line bisecting the bis-ischial line at right angles;

Femora;

Expulsion of meconium—not diagnostic. It may occur in cephalic births.

Identify foot, knee, shoulder, elbow, hand, by their anatomical characters;

In differentiating between head and breech, beware of relying on mere casual touch. Search minutely every accessible part of the presenting pole, and with firm pressure if impacted in the excavation.

Treatment.—*Before labor.* External version is permissible when it can be done without violence.

During labor. The danger to the child is chiefly due to the difficulty of delivering the after-coming head before the child dies from compression of the funis. Undelivered, the child dies within five minutes after the head engages and the utero-placental circulation is cut off.

The delivery of the after-coming head is facilitated by (1) full dilatation of the passages; (2) complete flexion

14

of the head, which also maintains flexion of the arms.

Promote 1 by preserving the membranes till they reach the pelvic floor, and, as a rule, keeping the breech intact. Accomplish 2 by avoiding traction till the trunk is delivered ; or, when traction is unavoidable, by external manipulation applied by an assistant.

A. Dorso-anterior positions. Preliminaries. Have the forceps ready. On expulsion of the trunk wrap the child's body in a flannel or towel to prevent premature efforts at respiration. Examine the funis for pulsation, and watch it for warning of danger to child. Pull the funis down and dispose by the sacro-iliac joint in that half of the pelvis which offers the most room.

Management of the arms:

a. Arms flexed: Deliver by the hand passed along the child's abdomen.

b. Arms extended: 1. *Delivery of the first arm.* As soon as a shoulder-blade can be reached, seize the feet and draw the trunk to the side opposite the occiput. Deliver the posterior arm first. Pass the free hand up along the dorsum and slip one or two fingers over the shoulder and along the humerus to the elbow. Sweep the elbow across the face and down.

2. *Delivery of the second arm.* Bring the child's trunk into the long axis of the mother's body. Seize the trunk with both hands, and push it up to unlock the head and extended arm from the grasp of the pelvic brim. Rotate the trunk to carry the undelivered arm opposite the nearest sacro-iliac joint. Assist rotation if necessary by

---

[1] Don't unnecessarily decompose the dilating wedge by pulling down the feet.

traction upon the delivered arm, drawing it across the child's back. Then, changing hands, deliver the second arm in the same manner as the first. Or, while sweeping the arm across the face, assist rotation of the head by external pressure on the occiput.

Delivery of the after-coming head. A. Dorso-anterior positions. Rotate the head to bring the face opposite one of the sacro-iliac joints. Combine: 1. *Expressio fœtus* by an assistant. 2. All the voluntary expelling powers of the mother. 3. The Smellie-Veit (Mauriceau) method, as follows :

Apply traction by two fingers hooked over the shoulders, astride the neck, maintaining extreme flexion by two fingers of the other hand pressed against the fossæ caninæ or the lower maxilla, the trunk lying along the forearm of the operator or upon the abdomen of the mother. Keep the long diameter of the head in the oblique diameter of the pelvis until past the brim. As the chin approaches the vulvar orifice, place two fingers in the child's mouth and depress the tongue for the admission of air ; the time thus gained may save the child's life. If it begins to breathe, time may be taken for the gradual delivery of the head, and the danger of perineal injuries be thus diminished.

Wigand-Martin method. This is the most effectual manual method if compelled to operate without assistance. The technique is simple. Flexion is maintained by the fingers of one hand in the child's mouth, or over the fossæ caninæ, and the head is driven down through the pelvis by suprapubic pressure with the other hand.

Forceps. Let an assistant hold the child's trunk well up on the abdomen of the mother. The forceps is then applied to the head and the delivery completed. This is the most certain of all methods of delivering the after-coming head. With care to observe the mechanism and to avoid violence, the danger of maternal injuries is no greater than in manual extraction.

B. Dorso-posterior positions. Rotate the occiput to the front after the expulsion of the body by gentle torsion of the trunk, with the aid of external pressure applied over the mother's abdomen by an assistant. Then proceed as in primary anterior positions. Rotation failing, deliver, by traction and suprapubic pressure, downward and backward over the perineum. The chin catching over the brim of the pelvis, deliver, occiput first, by traction upon the body upward and forward over the pubes, aided by suprapubic pressure.

Nuchal arm. On delivery of the trunk rotate the body in the direction *from* the misplaced arm, guarding against much torsion of the neck. Rotation more than a quarter circle endangers the spinal cord. The rotation of the head may be assisted by external pressure. Sometimes the nuchal arm may be best dislodged by the hand in the passages. The arms disengaged, proceed as in ordinary cases.

In failure of the powers at or above the brim, bring down one or both feet, if this be possible without violence.

Impaction. Deliver by traction with finger, fillet, or forceps. In dorso-anterior positions bring down the breech with the finger or fillet in one groin.

In dorso-posterior positions adjust a soft oiled fillet so that the loop may encircle the pelvis, the free ends depending between the thighs; or pass the fillet over one groin and hold it in place with one hand, making careful traction with the other.

Forceps. In cases not manageable by the finger or the fillet, apply forceps to the breech. Place one blade over the sacrum and ilium, the other over the posterior surface of the opposite thigh, or adjust the blades over the trochanters, avoiding pressure upon the ilia. Use moderate traction, assisted by *expressio fœtus*.

The cephalotribe is permissible on the dead fetus.

TRANSVERSE PRESENTATION. SHOULDER PRESENTATION.

A transverse presentation is one in which the long axis of the fetus lies across the long axis of the uterus. The presentation is, in fact, oblique rather than transverse. In a considerable proportion of cases cross presentations are spontaneously converted into longitudinal before labor begins. In persistent transverse presentation the shoulder becomes the presenting part after labor is established.

Frequency.—1 : 250.

Causes.—Unusual mobility of the fetus; twin pregnancy; fetal tumors; myomata of the lower uterine segment; undue pelvic inclination; pelvic deformity; low attachment of the placenta.

Positions:

       Left scapula-anterior—L. Sc.-A.

       Right scapula-anterior—R. Sc.-A.

       Right scapula-posterior—R. Sc.-P.

       Left scapula-posterior—L. Sc.-P.

It will be seen that these positions are named according to the situation of the presenting scapula. Scapula to the left and front is a left scapula-anterior; to the right and front a right scapula-anterior, and so on.

Prognosis.—In persistent transverse positions one out of ten of the mothers and one-half of the children die. The dangers to the mother are pressure effects, exhaustion, rupture of the uterus; to the child, pressure effects, prolapsus funis.

Diagnosis.—*Abdominal signs:*

> Absence of both fetal poles from the excavation during labor;
> Presence of the head in one iliac fossa.

Vaginal signs:

> Glove-finger protrusion of the membranes;
> Absence of the hard globular head;
> Absence of any presenting part at the beginning of labor;
> Presenting part a small rounded prominence after labor is established; distinguished from the ischial tuberosity by the absence of a companion; from it radiate the humerus, the clavicle, the spine of the scapula;
> Neck on one side of the presenting part, ribs on the other;
> Axilla;
> Elbow, identified by the olecranon.

The **diagnosis of position** is made by the location of the head, right or left, and of the scapula, anteriorly or posteriorly. The axilla or elbow looks toward the feet; the thumb toward the head.

Distinguish hand from foot, right from left hand. Shake hands with the fetus—the right hand of the examiner fits the right hand of the fetus, and conversely.

Spontaneous Delivery.

Spontaneous version. Rarely a shoulder presentation is converted into a breech, still more rarely into a vertex birth during labor.

Spontaneous evolution. The breech is forced past the presenting shoulder and delivered first.

Expulsion with trunk doubled on itself is possible only when disproportion between size of the pelvis and fetus favors. It is almost invariably fatal to the child.

Treatment.—*Before labor.* Correct the presentation by external cephalic version. Retain by an abdominal binder and lateral compresses.

During labor. In ordinary cases preserve the membranes; evacuate the bladder and rectum; note capacity of pelvis, size of child, situation of retraction ring and degree of thinning of lower uterine segment. Perform version, cephalic or podalic, by the external, bi-polar, or internal method, under anesthesia. Reduction of the malpresentation is often possible by the aid of the genu-pectoral position, without anesthesia.

In impacted and irreducible cases decapitation will be required.

TREATMENT OF COMPLEX PRESENTATIONS.

Head and hand. If possible, replace the hand; this failing, deliver by forceps, placing the arm in the unoccupied side of the pelvis, or perform podalic version.

Hand and foot or *head, hand and foot.* Bring down one or both feet.

Nuchal arm. Diagnosis is made by the hand in the passages, under chloroform.

Dislodge the arm with the hand in the uterus by rotating the body *from* the nuchal arm. Rarely version may be necessary.

In most cases of complex presentation, if the fetus is dead, delivery is best accomplished by craniotomy, in the interest of the mother.

Anomalies of Fetal Development.

TWINS.

Relative situations of twins are : one beside the other, one above the other, or one in front of the other.

Diagnosis of twins.

a. *Abdominal signs.* Excessive size and tension of the tumor; permanent tension of tumor with very limited mobility of contents should suggest twins;

Shape of the tumor—excessive width ; longitudinal sulcus—not diagnostic ;

Supra-pubic edema—occurs also in simple hydramnios;

Multitude of small parts ;

Two dorsal planes ;

Three or four fetal poles ;

One head in the excavation and one in the upper uterine segment ;

One head in the excavation and one in the iliac fossa ;

Distance from pelvic pole to fundal pole over thirty and a half centimetres (twelve inches).

Two fetal heart-sounds of different rates.

Two fetal heart-sounds of the same rate, but in widely different situations and on opposite sides ;

Heart above the umbilicus and head in the excavation.

b. Vaginal signs. Rapidly successive presentation of a head and a breech ;

Four extremities presenting ;

Two amniotic bags at the cervix.

Management of labor in twin births. The management of labor in twin births does not differ essentially from that of ordinary labor. Since the passages are prepared by the birth of the first child, the second birth is usually rapid or may be safely made so.

INTERLOCKING TWINS.

Presentations. a. Both cephalic. Both heads offering, one impacted between the head and trunk of the other fetus.

b. One cephalic, one pelvic. The after-coming head of the breech birth impacted between the head and trunk of the other fetus.

Management. Disengage by combining internal and external manipulation with the aid of the knee-chest position. Decapitate the first child as a last resort.

DOUBLE MONSTERS.

Premature birth and spontaneous delivery is the rule. Embryotomy may be required in difficult cases.

HYDROCEPHALUS

is an enlargement of the cranial vault, due to serous effusion into the cranial cavity. The effusion is usually found in the ventricles; very rarely in the arachnoid sac. The quantity of fluid may amount to several pints. Spina bifida or other anomalies generally coexist. The etiology is obscure.

Diagnosis.

a. Head-first cases.

Abdominal signs. Measurements of the head taken with a pelvimeter or calipers through the abdominal wall, or estimated by palpation.

Vaginal signs:

Size of the cranial vault;

Elasticity of the cranial vault;

Fluctuation of the cranial vault;

Increased width of sutures; this, however, may be present in the absence of hydrocephalus;

Large fontanelles;

Prominence of the frontal and parietal bones;

Sometimes a supplementary fontanelle between the anterior and posterior.

Confirm the diagnosis, if necessary, with the hand in the vagina, under anesthesia.

b. Head-last cases. One in five presents by the breech. The signs are:

Body wasted;

Head arrested after birth of the trunk;

Measurement or palpation of the head through abdominal wall.

Prognosis.—*Child.* Mortality over 80 per cent.; if born alive, the viability is feeble; nearly all die soon after birth.

Mother. Mortality 18 per cent., from exhaustion, rupture of the uterus, hemorrhage.

Treatment.—Delivery may be left to Nature or may be accomplished by version or perforation, according to the degree of obstruction. Aspiration of the cavity with a small trocar passed through a fontanelle or suture may frequently be substituted for craniotomy. The life of the child is not necessarily sacrificed by drawing off the fluid. The cephalotribe may be used as a tractor after perforation.

In difficult head-last cases, perforate the head, or open the spinal canal and catheterize the cranial cavity. The perforator may be safely passed beneath the skin, entering it over the neck below the vulva.

SEROUS EFFUSIONS INTO OTHER CAVITIES,

in case of marked dystocia, are to be managed by aspiration of the dropsical cavities, or by incision.

TUMORS.

Hygroma, fibroma, carcinoma, spina bifida, enlargement of abdominal viscera, and other tumors, are occasionally met with.

Treatment. When delivery of the fetus intact is impossible, fluid tumors may be reduced by tapping or incision; solid, by segmentation.

ANOMALIES OF LABOR ARISING FROM ACCIDENTS OR DISEASE.

PROLAPSUS FUNIS.

In this accident a loop of the cord slips down in advance of the presenting part of the fetus. As the labor goes on, the prolapsed portion of the cord is compressed between the part presenting and the walls of the birth canal, and without relief the fetus dies within two or three minutes from interruption of the feto-placental circulation.

Frequency.—It occurs once in about 250 labors.

Causes.—Anything which prevents the presenting part from completely and continuously filling the lower uterine segment, *e. g.*:

> Hydramnios;
> Pelvic deformity;
> Malpresentation; (frequency in head presentation, $1 : 304$; face, $1 : 32$; pelvic, $1 : 21$; shoulder, $1 : 12$);
> Complex presentations;
> Twins;
> Pendulous abdomen;
> Uterine myomata;
> Low placental insertion; also
> Marginal insertion of the cord;
> Excessive length of the cord.

Prognosis.—There is no increased risk to the mother from the prolapse itself; there may be from the conditions which give rise to it.

The fetal mortality is 50 per cent. It is greatest in vertex presentations and in first labors. The danger is obviously increased after the rupture of the membranes.

Diagnosis.—Before rupture of the membranes, differentiate from fingers and toes by their anatomical characters. The fetal parts will often be drawn up out of the way when touched. In rupture of the uterus distinguish from a loop of intestine. Examine the prolapsed part for the funic pulse.

Continuous absence of pulsation in the cord for fifteen minutes may be taken as evidence of the death of the fetus. Examine for the fetal heart over the abdomen.

Treatment.

Before rupture of the membranes. Preserve the membranes. They should in no case be intentionally ruptured without first examining for possible prolapse of the cord. Maintain the latero-prone position. Place the patient on the side opposite that on which the prolapsed cord lies, in the hope that the displaced loop may return by its own weight. Push the cord up carefully between the pains, avoiding rupture of the membranes. Guard against a recurrence of the prolapse until the presenting part has fully engaged. Listen frequently over the abdomen for the fetal heart

After rupture of the membranes. Replace at once if the funic pulse can be felt; if pulsation has ceased and the heart is still beating, push up the presenting pole and reposit the cord after pulsation returns.

a. Manual method of repositing. Place the patient

in the latero-prone or the genu-pectoral posture. Twist
the prolapsed loop lightly into a rope and replace anteriorly by taxis, between the pains. Much handling of
the cord is dangerous—it enfeebles the fetal heart. For
retention, crowd the presenting pole firmly into the
pelvic brim, and hold it there by manual pressure or by
an abdominal binder. Keep the patient in the latero-
prone position with the hips elevated. Examine *per
vaginam* from time to time, lest the cord again slip down
as the labor advances.

Listen at short intervals for the frequency and quality
of the fetal pulse. Do not subject the mother to the
manipulation for reposition if the child is surely dead or
non-viable.

b. Instrumental method of repositing. Posture as in
the manual method. Reposit by means of an English
catheter with a tape attached and loosely looped over
the cord. After complete reposition leave the catheter
in the uterus. The instrument should be armed with a
stylet, which should be withdrawn after repositing.
Use measures for retention as in the manual method.

c. Should all attempts at reposition and retention fail,
it is sometimes possible to save the child by rapid
delivery. This may be done in vertex presentation by
the forceps or by version ; in breech cases by the usual
method of breech extraction. Version is sometimes
better primarily than attempts at reposition.

INVERSION OF THE UTERUS.

Inversion of the uterus may be partial or complete.
It generally begins as a cup-shaped depression of the
fundus ; rarely it begins at the cervix.

Frequency.—About 1 : 200,000.

Etiology.—Inertia uteri in the third stage is the principal cause. Maladroit pressure on the fundus, traction on the cord while the uterus is relaxed, or a fundal placental seat may be complicating causes. Inversion of the uterus rarely occurs after delivery of the placenta.

Prognosis.—Grave without prompt reposition. The mortality, even under skilful management, is one-fifth to one-third, from hemorrhage and shock or peritonitis and gangrene of the uterus.

Diagnostic Signs.—Inversion of the uterus is attended with shock, pain, hemorrhage, and vesical and rectal tenesmus.

In partial inversion beginning in the upper segment a cup-like depression can be felt at the fundus by abdominal touch.

For the diagnosis of complete inversion, catheterize the bladder and empty the rectum before examination. Exclude morbid growths. The condition is recognized by the absence of the usual abdominal tumor, as demonstrated by abdominal palpation or by combined abdominal and vaginal or rectal examination; by the presence of a vaginal tumor; and by the character of the tumor.

The inverted uterus is distinguished from a polypus by special contractility, by its large pedicle, and by pain and immobility on attempting torsion. In case of polypus depending from the cervix, a sound may be passed alongside the tumor into the uterine cavity. Differentiation is sometimes difficult. Note that the placenta may still be adherent.

Treatment.—The preventive treatment consists in the proper management of the third stage of labor.

Methods of Reposition.

a. Simple cases (within a few hours after inversion). With the patient under an anesthetic, the operator places one hand on the abdomen over the inverted uterus for counter-pressure, cones the fingers of the other hand, and, passing it into the vagina, applies the finger tips over the insertion of one Fallopian tube. The uterus once fairly indented at this point, complete reduction is easy. The force should be directed to one side of the sacral promontory. If the placenta is adherent, replace all; if partially detached, separate and remove it before replacing.

b. Difficult cases. Apply taxis with the aid of the genu-pectoral or latero-prone position. This failing, recourse should be had to elastic pressure by means of a water-bag, alternated with taxis. Leave the water-bag in place for six or eight hours, then remove it and try taxis. Failing replace the water-bag for another six hours. The utmost care must be observed to prevent sepsis. Extreme measures should be avoided during the puerperium, and attempts at reposition should be postponed for two or three weeks if not successful within twenty-four or forty-eight hours.

RUPTURE OF THE UTERUS.

Nature of the Accident.—The tear generally begins in the lower segment. It may take any direction and go to any extent within the limits of the organ; it may invade the vagina and the bladder. The portio vaginalis may be torn off. Notable fissures of the cervix occur in nearly all labors. The rupture is complete when it extends from the uterine into the peritoneal cavity; otherwise it is incomplete. Spontaneous rupture occurs rarely during pregnancy, most frequently near the close of the first stage of labor.

Frequency.—1 : 4000.

Causes.

a. Predisposing. Anything which impairs the integrity of the uterine muscularis, as carcinoma, myoma, the cicatrix of a former laceration or of a Cesarean section not properly sutured. Obstructed labor, leading to excessive thinning of the lower uterine segment, is the chief predisposing cause.

b. Exciting. Ergot; operative violence, such as forceps through an undilated os, version in a contracted uterus, etc.

Prognosis.—The maternal mortality is 90 to 95 per cent., from hemorrhage, peritonitis, or septicemia. The fetal mortality is still greater.

Diagnosis.

Precursory signs:

Concurrence of obstruction with violent uterine effort;

Excessive uterine retraction—locate the retraction ring by abdominal palpation.

Signs of rupture:

Collapse;

Hemorrhage—external, sub-peritoneal, intra-peritoneal;

Persistent local pain;

Sudden cessation of labor pains—in complete rupture;

15

Sensation of tearing;

Recession of the presenting part;

Absence of the signs of fetal life;

Prolapse of intestines into the uterus;

Uterus and child presenting separate tumors.

The diagnosis should be confirmed by exploration with the fingers in the uterus.

Treatment.

1. *Preventive.* Remove the cause of obstruction if possible; correct malpositions. Promptly resort to artificial delivery if indicated by excessive retraction of the uterus.

2. *Remedial.*

a. The fetus should be immediately delivered by the natural passages if it is still wholly or mainly in the uterus. In vertex cases perforation should be done in the grasp of the cephalotribe or forceps, if demanded in the interest of the mother, and especially if the fetus be dead. The placenta should be removed. If it has escaped into the peritoneum it may be drawn down to the uterine wound by traction on the cord, and delivered by the hand. Prolapsed intestines must be replaced. Drainage should then be established as follows: Fold a large rubber tube, tie the limbs of the tube together, make several perforations in the bight of the tube and pass the bight up through the uterine rent and about an inch beyond; or drain in similar manner with iodoform wicking or gauze. Keep the uterus contracted. Remove the drain in three to five days, on cessation of notable discharge.

b. Celiotomy should be done in case of fetus wholly in the peritoneal cavity, fetus long dead, much hemorrhage into peritoneum, cervix not dilatable, or site of rupture unfavorable for drainage. Close the uterine lacerations by deep suture. Cleanse the peritoneum by irrigation with the normal salt solution.

Amputation of the uterus may be resorted to if necessary to avert sepsis; especially is this advisable in extensive lacerations or in case of an infected uterus. The method may be the same as in the Porro operation, or the stump may be treated in accordance with modern extra-peritoneal methods practised in amputations for fibroids.

THE HEMORRHAGES.

ANTE-PARTUM HEMORRHAGE.

1. *Placenta Previa.*

Definition.—The placenta is previa when its site encroaches upon that part of the uterus which is concerned in the dilatation of the first stage of labor.

Degrees of placenta previa:

1. Marginal—edge encroaching upon the zone of dilatation.
2. Partial—partially covering the fully dilated os uteri.
3. Complete—wholly covering the fully dilated os. Exact central implantation is rare.

Frequency.—1 : 1000. More frequent in multiparæ than in primiparæ.

Causes.—Conditions leading to tardy fixation of the ovum, permitting it to fall into the lower uterine segment, such as:

Morbid conditions of the uterine mucosa;
Enlargement of the uterus;
Relaxation of the uterus.

The cause of hemorrhage is the partial separation of the placenta during canalization of the cervix.

The source of the hemorrhage is the uterus, sometimes the placenta as well.

Prognosis.—*Mother*. Mortality, in cases that go to the later months, is one-fifth to one-fourth, including deaths from the sequelæ. Two-thirds of the children die.

Mortality to both mother and child varies, however, with the degree of placenta previa. The maternal mortality is due to hemorrhage, shock, sepsis, and thrombotic affections; the fetal, to apnea, hemorrhage, prematurity, operative causes. There are practically no maternal deaths from placenta previa before the seventh month. The danger increases as gestation advances, owing to the increasing size of the vessels and the progressive loosening of the placental attachment.

Diagnosis.

Symptoms. Usually none in the early months. The first indication generally is a sudden hemorrhage of greater or less severity. The first hemorrhage occurs most frequently in the seventh or eighth month, sometimes not till term. Hemorrhage of any note during pregnancy demands investigation, especially in the later months. Bleeding from placenta previa during labor is most profuse in the intervals between the pains.

Physical Signs.

a. Abdominal. The placenta may sometimes be mapped out by abdominal palpation. Under the placenta the fetal parts are obscure, elsewhere they are more distinctly felt. The convex edge of the placenta may often be traced as a resisting ring.

b. *Vaginal:*

> Unusual development of the cervix, especially in complete placenta previa ;
>
> A boggy feel of the cervix and the lower segment of the uterus ;
>
> A cushiony mass between the presenting part and the examining finger ;
>
> Characteristic stringy feel of the detached surface of the placenta, examined through the cervical canal ; distinguish from clots which are more friable.
>
> In marginal placenta previa the edge may be felt if separated.

Treatment.

a. Before viability. In general, expectant. If the hemorrhage be copious, placenta previa complete, or the fetus dead, empty the uterus.

b. After viability. Induction of labor, simple cases excepted.

. *Management of labor.* The principal indication is the control of hemorrhage. The hemorrhage controlled, wait, but remain with the patient till delivered. Nature is competent in very rare cases by extra-rapid delivery.

Rupture of the membranes and the application of a firm binder may suffice in marginal and in certain cases of partial placenta previa with but little hemorrhage. The presenting pole serves as a tampon.

After sufficient dilatation forceps, with very moderate traction, is permissible in similar conditions to hold the head in the lower uterine segment, as a tampon.

The vaginal tamponade should be mainly restricted to cases in which there is little or no dilatation of the cervix. Material may be the usual cotton pledgets, or gauze—in strips—impregnated with zinc oxide and sterilized. The vagina should be sterilized before placing the tampon. Remove in six or eight hours. Renewal of the tamponade is rarely required. Barnes' bags in the cervix may be of service in similar cases for the double purpose of controlling hemorrhage and promoting dilatation.

Podalic version is an effectual measure for controlling the hemorrhage. It is especially indicated in case of much bleeding. With one or both feet down, the fetus acts as a conical cervical plug. Bi-polar version has the advantage that it may be done as soon as one or two fingers can be passed through the cervix. External version, when practicable, may be done earlier. In the bi-polar or internal method the edge of the placenta is pushed aside and fingers passed through the membranes. Even after sufficient dilatation it will rarely be necessary to pass the entire hand into the uterus. After version extract very slowly and with extreme care, to avoid shock, or leave the expulsion to Nature.

Other methods. Separation of the placenta from the lower uterine segment (Barnes), permits retraction of the zone uncovered. The area of detachment should be not less than four and a half inches in diameter.

Complete separation and extraction of the placenta (Simpson) is applicable in case the child is dead or not yet viable.

Extraction of the child by perforation of the placenta is rarely permissible.

Precautions. Avoid too precipitate and violent interference. It is the cause of a large proportion of deaths in placenta previa. Guard against shock, septic infection, post-partum hemorrhage. Give ergot for several days after labor.

2. *Accidental Hemorrhage.*

This term is applied to bleeding caused by partial separation of a normally situated placenta, and occurring in the later months of pregnancy.

Varieties.

a. Apparent, in which the blood escapes by the vagina.

b. Concealed, in which the effused blood accumulates in the uterine cavity. Either of the following conditions may be present.

1. Placenta separated at the center, margin adherent.

2. Placenta separated at the margin partially lifting the membranes beyond the margin.

3. Separation the same as in 2, but blood escaping into the cavity of the ovum by rupture of the overlying membranes.

4. Separation of the margin of the placenta and of the membranes, but the lower segment of the uterus blocked by the fetal head.

Causes: The loose attachment of the placenta, normal to the last weeks of gestation;

Violent muscular exertion;

Violent uterine contractions;

External violence;

Blood state, *e. g.,* of albuminuria, anemia;

Placental disease.

Prognosis.

Apparent variety. Not usually grave for the mother, frequently fatal to the child.

Concealed variety.   The maternal mortality is fifty per cent., from shock due to hyperdistention of the uterus or operative violence, from blood-loss, post-partum hemorrhage, or sequelæ; the fetal, ninety per cent. or more.

Diagnosis.

Apparent variety.   Distinguish from rupture of the uterus, which occurs later in labor and is attended with recession of the presenting part, with diminution of the uterine tumor and the development of a new abdominal tumor; from placenta previa by absence of the physical signs of misplaced placenta.

Concealed.   The principal signs are:

Uterine distention;

A node or boss on the surface of the uterus at the site of the retro-placental blood collection;

Atony of the uterus;

Uterine tumor doughy;

Fetal parts obscured;

Continuous pain in certain cases from distention of the peritoneal coat of the uterus;

Bloody liquor amnii—push up the presenting part and allow a portion of the liquor amnii to escape to see if it is bloody;

Fetal heart tones feeble, irregular;

Signs of internal hemorrhage, viz.:

Collapse;

Pallor;

Surface cold, clammy, especially the extremities;

Perspiration;

Respiration irregular, sighing, sobbing, yawning;

Pulse rapid, thready, compressible;

Thirst;

Jactitation ;

Tinnitus aurium ;

Dyspnea ;

Nausea ;

Dimness of vision ;

Syncope.

Concealed and slight apparent hemorrhage may coexist.

Treatment.—In either variety dilate the cervix, manually, and rupture the membranes. Maintain firm compression of the uterus by means of a binder, and the use of ergot hypodermatically. After full dilatation deliver by forceps or version, or in dead or non-viable fetus by craniotomy. Be prepared for post-partum hemorrhage.

Treatment of Acute Anemia.—Elevate the hips and lower the head. Bandage the extremities—autotransfusion—temporarily. Make hot applications to the feet. Give opium in full doses, gr. ij, p. r. n., or its equivalent, to fill the cerebral vessels, hypodermatic injections of brandy or ether, fluid extract of digitalis, $\mathfrak{m}$ j to $\mathfrak{m}$ v, strychninæ sulph., gr. $\frac{1}{30}$, trinitrin, gr. $\frac{1}{100}$ to $\frac{1}{50}$ or $\frac{1}{25}$ repeated p. r. n. The injection of the normal salt solution (six-tenths of one per cent.; approximately, gr. iij ad ℥j), into the rectum, or hypodermatically, between the scapulæ, or into a vein, is a measure of great value. The salt solution should be previously sterilized by boiling, and should be injected at the temperature of 100° F. From one to three pints may be used.

An easily improvised apparatus for intravenous infusion may be made with a glass funnel, two feet of rubber tubing and a glass or metal canula. All should be sterilized by boiling.

Keep the large bowel filled with the saline solution, with plain warm water, or suitable nutrient enemata.

For the thirst, give a saline drink—*e. g.*, a weak solution of ammon. acetat. Fluids by the stomach must be fed in small quantities and often, beginning with ℥j, at intervals of one or two minutes. Plain hot water, brandy, or whiskey and water are useful restoratives. Nutrient fluids may be given after a few hours.

POST-PARTUM HEMORRHAGE.

The term post-partum hemorrhage is applied to hemorrhage which occurs immediately after the birth of the child and has its origin in the placental site. The accident is rare in well-managed labors. Hemorrhage from lacerations of the passages is not strictly within the meaning of this term.

The normal blood loss at the close of labor varies from two or three ounces to a pint.

Causes.—Failure of ligation of the uterine vessels owing to inertia uteri from exhaustion and previous overdistention of the uterus, mismanaged third stage, excess of chloroform, full bladder or rectum. The retention of blood coagula or of portions of the membranes or placenta may prevent full retraction and thus interfere with the closure of the vessels. Uterine neoplasms may act in like manner.

Diagnosis.

Danger signals :

    History of hemorrhage in previous labors ;

    Pulse at or above 100 ;

Imperfect retraction — detected by abdominal palpation ;

Presence of other causes of hemorrhage.

Signs.

A sudden outburst of blood ;

No uterine globe ;

Systemic effects of severe hemorrhage. (See page 232.)

The absence of visible flooding does not forbid the diagnosis of hemorrhage. A flow of blood with firm uterine contraction does not come from the uterine cavity. Examine for cervical or vaginal lacerations.

Treatment.—*Prophylaxis.* Measures for prophylaxis must be addressed mainly to the uterine retraction. Watch the uterus, with the hand continuously on the abdomen, from the beginning of the third stage and for a half-hour or more after the placenta is delivered. Use friction if required to promote normal contractions. Extract. ergot. fld., ℥ss, hypodermatically, is a valuable prophylactic. It is especially indicated after the use of chloroform, and in other conditions which predispose to hemorrhage. It may be used at any time after the birth of the head when hemorrhage threatens.

Remedial measures. a. Simple cases. Manipulation, with one or both hands, over the abdomen; fluid extract of ergot, ℥j to ℥jss hypodermatically ; hot intra-uterine douche, at a temperature of 110° to 120° F.

b. Severe cases. Combined internal and external pressure and kneading of the fundus with one hand in the uterus and the other over the abdomen; hot intra-uterine douche at a temperature of 115° to 125° F. ;

hand in the uterus, raking the cavity vigorously with the finger tips.

c. Uterine tamponade. The most effective measure for the control of severe post-partum hemorrhage is the uterine tamponade with iodoform gauze or simple sterilized gauze. It should be reserved, however, for desperate cases.

Method. Seize the cervix with a volsella, and draw it well down. Push the gauze into the cavity of the uterus with a uterine dressing forceps, or with the fingers. In the absence of instruments the gauze may be packed with the fingers alone. Remove cautiously in from twelve to twenty-four hours.

Additional measures. Child to the breast as a reflex excito-motor ; compression of the aorta, very effectual as a temporary expedient ; flagellation of the lower part of the abdomen with a wet towel ; faradism of the uterus, one pole within the uterus and one over the abdomen or upper sacral region, or both poles over the abdomen, one on either side of the uterus ; or the use of the curette.

Hemorrhage from a torn cervix is best controlled by suture. The first stitch should be passed above the angle of the tear. Vaginal hemorrhage may be managed by pressure, better by suture. Treat the anemia as in other cases.

SECONDARY POST-PARTUM HEMORRHAGE.

Definition.—A hemorrhage from the placental site occurring within the post-partum month later than six hours after labor.

Causes.—Retention of membranes or placental fragments; congestion of the uterus from misplacement or other causes; too early getting up ; violent emotion.

Treatment.—Remove the causes ; keep the patient in bed; correct misplacements of the uterus. Give hot vaginal douches, two or three gallons, temperature 115° to 120° F., morning and evening; this failing, curette the uterine cavity and pack with iodoform gauze; remove the packing in forty-eight hours.

SEPARATION OF THE SYMPHYSIS PUBIS

may occur spontaneously from excessive relaxation of the pelvic joints at term ; more frequently it is the result of unskilful use of forceps. The vagina and bladder may be lacerated.

Diagnostic Signs.—Mobility of the ends of the pubic bones upon each other ; sulcus between the bones; later, outward rotation of the femora, locomotion impeded.

Treatment.—Support by means of a firm pelvic bandage maintained for at least four weeks. Neglected cases may be treated by vivifying the joint surfaces and suturing the bones with silkworm-gut, catgut, or silver wire.

ECLAMPSIA.

Definition.—The term puerperal eclampsia is synonymous with puerperal convulsions. The convulsions are epileptiform in character, are usually associated with albuminuria, and may occur in advanced pregnancy or in the first few days of the puerperium. Con-

vulsive attacks in childbed from hysteria, epilepsy or
cerebral lesions are not included under this term.

Frequency.—Eclampsia is met with in 1 : 500 cases
of advanced gestation ; 1 : 4 cases of pregnancy nephritis. Nephritis is found in 6 per cent. of gravid
women that go to term. It is most frequent in primiparæ and in multiple pregnancies, and occurs more
commonly in pregnancy or labor than in the puerperal
period.

Etiology.—The principal cause of the convulsions is a
toxemia due to imperfect elimination by the kidneys—
uremia. Reflex irritation from the uterus is a potent
co-operating cause. Rarely, the renal lesion is nothing
more than acute insufficiency, generally it is an acute
parenchymatous nephritis. Sometimes the acute nephritis is engrafted upon a chronic. The cause of the
nephritis of pregnancy is obscure. According to Tyson,
it is due to the irritating effect of a toxic substance in
the maternal blood contributed by both mother and
fetus.

Premonitory Symptoms and Signs.

> Albuminuria ;
> Tube-casts in the urine ;
> Edema, especially of the face ;
> Debility ;
> Headache, generally frontal, sub-occipital rarely ;
> Nausea and other digestive disorders ;
> Contracted pupils ;
> Visual disturbances ;
> Epigastric pain.

Differential Diagnosis.—Distinguish from hysterical
and epileptic convulsions by the urinary examination.

Clinical Phenomena.—The attack is usually preceded by the symptoms already referred to. At the onset of the convulsion the eyes become fixed. Spasmodic movements begin in the facial muscles, then become general. The convulsion is at first clonic, then tonic. The patient is asphyxiated from tonic spasm of the respiratory muscles. Froth, generally bloody, flows from the mouth and nostrils.

The duration of the convulsion is one or two minutes; the interval between the attacks varies from a few minutes to several hours.

Coma follows the eclamptic seizure, usually subsiding within half an hour. The pulse ranges from 100 to 140. The temperature varies in different cases from normal, or subnormal, to 105° F. or more.

Prognosis.—The prognosis is graver the earlier the attack in pregnancy or labor. The danger increases with the number of seizures. Recovery is rare after fifteen or twenty convulsions; seldom occurs after a temperature of 105° F. Impairment of the mental faculties sometimes remains.

The nephritis of pregnancy in women pregnant for the first time after forty years of age is uniformly fatal if the pregnancy is allowed to go to the later months. (Tyson.)

Pregnancy in primiparæ having nephritis before conception is invariably fatal if not interrupted before term. (Tyson.)

The maternal mortality is about 25 per cent., from exhaustion, slow asphyxia during convulsion, cerebral hemorrhage, or edema of the lungs; fetal, at least 50 per cent., mainly from asphyxia.

Treatment.—*Prophylactic.* A milk diet limits the urea production, but is not always well borne, and does not alone serve for proper nutrition for much more than a week. Catharsis by salines and diaphoresis by hot baths, hot packs and the use of spts. ether. nitrosi helps by supplementing the crippled action of the kidneys. Hot alkaline drinks are suitable diuretics; hot fomentations and dry cups over the kidneys may be used for the same purpose. Extract. veratri viridis fl. (Squibb), ♏ iij to ♏ vj t. i. d., or enough to hold the pulse below 70, is a valuable prophylactic. Chloral, ℈j to ℈iij daily, or the bromides in similar doses are the most useful agents for suppressing the reflexes. Iron is usually required as a restorative. Pronounced uremic symptoms, a large proportion of albumin in the urine, or scanty urinary secretion, when not immediately relieved by the foregoing measures, call for the induction of labor.

Remedial. The principal reliance for controlling the convulsions is the combined use of chloroform, extract. veratri viridis fl., catharsis, and the prompt evacuation of the uterus. The veratrum viride may be replaced with chloral with equally good effect.

Pending the action of other remedies, place the patient immediately under chloroform nearly or quite to the surgical degree. Chloroform is an almost certain anti-eclamptic. Its use is imperative during operative interference. Prolonged chloroform narcosis, however, is dangerous; two or three hours should generally be the limit.

Inject subcutaneously ext. veratri viridis fl. (Squibb) ♏ x to ♏ xx. At the end of a half-hour inject ♏ x if the pulse is not then below 60. A convulsion is practically impossible while the patient is sufficiently under the influence of veratrum to hold the pulse-rate below 60. The patient must not be permitted to assume the upright posture while using the drug in large doses. Collapse under veratrum is readily combated by the use of morphine, hypodermatically, or by whiskey given in the same manner or by the bowel.

If chloral is preferred to veratrum, it is best administered by the rectum in a teacupful of milk. The dose should be ℨss repeated hourly till ℨj or ℨij have been given.

For catharsis, calomel and salines, elaterium (Merck) gr.¼, or ol. tiglii ♏j to ♏ij, may be used.

Labor usually sets in on the occurrence of convulsions. Measures are indicated to hasten the labor if already begun, or to induce it if not spontaneously established. The danger is substantially over in 90 per cent. of cases after delivery. For the induction of labor the most prompt and reliable method is the intra-uterine injection of glycerin (see page 167.) It is equally effective for accelerating the first stage. Recourse may be had to manual dilatation of the cervix, if necessary, after the os internum is fully obliterated. Other anti-eclamptic measures of repute are: morphine, gr. ½ to gr. 1½, hypodermatically; nitro-glycerin, gr. $\frac{1}{50}$, in similar manner; amyl nitrite, ♏v, by inhalation; the inhalation of oxygen; venesection. During convalescence iron and general tonics are indicated as restoratives.

DIABETES MELLITUS.

As has already been stated, sugar may be found in a large proportion of cases in the urine of women during the later weeks of pregnancy, and for a few days after childbirth. Generally, the glycosuria is due merely to resorption of lactose, and is unimportant. True diabetes mellitus is a dangerous complication of labor and the puerperal condition. Rarely it is a cause of fetal death.

CARDIAC DISEASE.

Valvular heart lesions are aggravated by the increased tax put upon the circulation in the later months. Sometimes they are the cause of abortion or premature labor.

Advanced cardiac lesions are a dangerous complication of labor. Engorgement of the right heart and edema of the lungs are liable to supervene. The danger is greatest at close of the third stage.

Treatment Before and During Labor.—Tr. strophanthi, ♏ v, q. v. h., or ext. digitalis fl., ♏ ij, guarded with trinitrin, gr. $\frac{1}{50}$ t. i. d.; strychninæ sulph., gr. $\frac{1}{40}$, t. i. d.; venesection in extreme venous engorgement; inhalation of amyl nitrite during third stage; ether should be used as the anesthetic, and only during the severe pains of labor. The use of ergot should be avoided, as a little extra blood-loss is conservative.

PATHOLOGY OF THE PUERPERIUM.

PUERPERAL INSANITY

May begin during pregnancy or the puerperium. In the puerperal period the onset occurs most frequently at the end of about two weeks, less frequently after five or six weeks.

Frequency.—About 1 : 400.

Causes. — Hereditary predisposition, bad mental hygiene, violent emotional disturbance, anemia, uremia, sepsis; distinguish from the transient delirium of septicemia.

Prognosis.—For life, generally favorable. The mortality does not exceed 5 per cent. Nearly 70 per cent. recover their reason.

Treatment. — Look to the mental and physical hygiene. In cases occurring during the puerperium, suspend nursing. Give iron, pil. Blaud, one or two t. i. d., or arseniate of iron, gr. $\frac{1}{10}$, t. i. d., in anemia. The hypodermatic use of the hydrobromate of hyoscine in doses of gr. $\frac{1}{200}$, two or three times daily, is a good sedative in maniacal forms. In melancholia, morphine, gr. $\frac{1}{8}$, once or twice daily, will be of service. Chloral or the bromides, gr.xx, or sulphonal in similar doses, are frequently useful as hypnotics. Treat infection as in other cases.

GALACTORRHEA.

By galactorrhea is meant an excessive secretion with constant oozing of milk from the nipples. Amount may reach several quarts daily. Quality thin and watery. May affect one or both breasts. Often results in severe, even dangerous impairment of the general health.

Treatment. Interruption of nursing. Compression of the breasts by means of a binder. Restriction of fluids. Potass. iodid., gr. v., t. i. d. Tonics and general restorative measures.

MAMMARY ABSCESS.

Frequency.—Occurs in 5 to 6 per cent. of nursing women.

Causes.—Predisposing causes are bad general health, lowering the resisting power; milk stasis, impairing the vitality of the epithelium of the lactiferous ducts; lesions of the nipples.

The exciting cause is sepsis.

Forms. — 1. Subcutaneous. 2. Glandular — parenchymatous mastitis—in the great majority of cases a lymphangitis. 3. Sub-glandular — paramastitis. Two or all forms may coexist.

Diagnosis.—*The subcutaneous form* presents the signs of ordinary phlegmon; is generally single.

The glandular form is characterized by more pain; more constitutional disturbance; generally ushered in by a chill; gland indurated; is often multiple.

The sub-glandular form. In this form the temperature is persistently high; the pain deep-seated; the gland not indurated, and it floats upon the underlying fluid. Pass an exploring needle beneath the gland.

Treatment.—1. *Prophylactic.* Massage is useful in simple milk engorgement—in the absence of inflammation. The breast should be gently stroked from the base toward the nipple. Limit the amount of fluids ingested. Relieve hypersecretion, if necessary, by saline cathartics, or in non-nursing patients by topical use of atropinæ oleas, and put the patient on tonics, especially quinine. The aseptic management and curative treatment of nipple lesions is of fundamental importance. Engorged breasts should be supported with a sling. An easily improvised support is the Murphy binder.[1] It should be so fitted as to act as a sling rather than a compress.

2. *Abortive.* Absolute rest of the gland, abstinence from liquids, a saline cathartic, oleate of atropia, locally, with care lest the milk secretion be too much repressed; quininæ sulph., gr. v bis die, are valuable abortive measures.

3. *Treatment of suppuration.* Open early and freely, with antiseptic precautions. The incision should radiate in a direction from the nipple, avoiding the areola. Thoroughly cleanse and disinfect the abscess cavity. For this purpose there is no better antiseptic than the peroxide of hydrogen. It is non-toxic and an efficient pus-destroyer. A counter-opening may be necessary for

---

[1] The Murphy binder is made of a straight piece of muslin with a deep notch cut in one side for each arm and one for the neck.

satisfactory drainage. Insert a drainage-tube, protect the wound with antiseptic dressings, and apply compression to keep the walls of the abscess cavity in contact.

Treatment of Sore Nipples.

Cleanse the nipples after each nursing with a saturated aqueous solution of salicylic or boric acid. Dry and pencil with fresh white of egg, or saturate with cocoa butter. The following nipple lotion is useful in excoriation: R. Plumbi nitrat. gr. x, glycerin. ℨij, aq. ad ℥j.

A good soothing and antiseptic dressing, and one which does not require washing off before nursing, is the following: R. Amyl. glycerit., bismuth. subnit., āā ℥ss. Cleanse with the salicylic solution after nursing and re-apply the bismuth mixture.

These measures failing, rest one nipple for twenty-four hours, or let the child nurse through a nipple shield.

For relief of pain during nursing, pencil five minutes before with a 1 per cent. solution of cocaine which has been sterilized by boiling.

Fissures may be dusted with powdered tannin, or lightly touched once daily with the solid stick of nitrate of silver, first pencilling with the cocaine solution ; or they may be painted with the compound tincture of benzoin several times daily.

PUERPERAL FEVER.   PUERPERAL SEPTICEMIA.

Frequency.—In pre-antiseptic times the usual mortality from childbed fever in hospitals was 2 to 6 per cent., and "epidemics" with a death-rate of 10 per cent. or even more were by no means infrequent. Now, in well-managed isolated maternities, less than $\frac{1}{2}$ of 1 per cent. of puerperal women die from puerperal infection

In general private practice, without antiseptics, there is little less than 1 per cent. of fatal cases. Under a strict asepsis there are practically no deaths from puerperal fever.

The disease occurs more frequently in primiparæ than in multiparæ.

Etiology.— *Cause,* septic infection of the birth canal; lowered resisting power favors.

The sources of infection are the lochia of puerperal-fever patients; secretions from suppurating wounds; erysipelas; diphtheria or scarlet fever in certain cases, owing to complications involving the presence of wound-infection germs; cadaveric and other dead and decomposing animal matter. Self-infection (auto-infection) in the strict sense of the term does not exist. The term as now used refers to infection from septic matter primarily present in the genital tract.

The vehicles of infection are the hands of the physician or nurse, instruments, utensils, cloths, germ-laden dust, etc.

The avenues of absorption are the obstetric wounds of the vulva, vagina, cervix, and corpus uteri, and even intact surfaces of the genital mucosa. Systemic infection springs most frequently from the uterine cavity, usually from the placental site.

The channels of diffusion are the lymphatics, and to some extent the veins.

Bacteriology. The organisms most frequently found are the streptococci—chain cocci; staphylococci are occasionally met with. The bacterium coli commune and certain other micro-organisms are occasional factors in the pathogeny. The rod-shaped bacteria of putrefaction are generally present. Putrefaction of lochia furnishes a favorable soil for the development of pathogenic organisms. The putrefactive bacteria act solely, others largely by their chemical products—ptomaines.

Possible Lesions are: Endometritis; salpingitis; ovaritis; metritis; parametritis; perimetritis or pelvic peritonitis; diffuse peritonitis; uterine lymphangitis and phlegmonous lymphadenitis—generally attended with peritonitis; phlebitis—uterine, para-uterine and crural; colpitis; pure septicemia—acute ptomaine poisoning—putrid intoxication; secondary affections, such as pneumonia, pleurisy, pericarditis, endocarditis, nephritis, arthritis, subcutaneous phlegmons, and others.

Prognosis.—As a rule, the earlier the attack the graver the prognosis. It is most unfavorable in acute putrid intoxication, diffuse purulent peritonitis, pyemia.

Diagnosis.

General Symptoms of Infection. The first symptoms are generally developed on the second or third day, rarely later than the fourth or fifth, since the obstetric wounds have by that time begun to granulate. The majority of cases begin insidiously. The attack is frequently ushered in by a chill or slight chilliness.

The most prominent early symptoms are a rise of the pulse (100 to 140), elevation of temperature (102° to

104° F.), fetid lochia—yet sepsis is possible without fetor. Exclude malarial pyrexia (by quinine), fecal retention, emotional, mammary and other non-septic causes.

Symptoms of Special Lesions.

Endometritis. After-pains severe and prolonged; cervix more patulous · than normal for the time; uterine lochia foul; bloody flow prolonged; involution retarded.

Parametritis and *perimetritis.* Localized pain and tenderness; moderate tympanites, frequently nausea, lochia scanty; exudate found in one or both sides of the pelvis on bimanual examination; uterus more or less fixed, sometimes displaced; fluctuation if pus forms—abscess forms in 20 per cent. of cases of parametritis.

Diffuse. peritonitis. Exquisite abdominal pain and tenderness in the early stages, usually; tympanites extreme; vomiting of greenish fluid, loose diarrheal discharges; later, collapse. Fatal cases terminate within a week, generally.

Phlegmasia alba dolens. a. Thrombo-phlebitic form. The lesions are venous thrombosis generally, phlebitis and peri-phlebitis. Develops several days and even weeks after delivery; is generally preceded by signs of pelvic inflammation or some form of sepsis; fever, first of a remittent, then of an intermittent type; pain in the affected limb; limb becomes swollen, tense, hard, white, glistening; affected veins may sometimes be felt on palpation as hard, irregular cords; resolution begins after about two weeks; duration may be many weeks; abscess formation or gangrene very rarely supervenes; there remains more or less edema on standing or walking, with impairment of muscular power, in a certain proportion of cases lasting months; recurring chills are a signal of metastatic affections; the disease may extend from one limb to the other.

b. Cellulitic form. Characterized by inflammation, suppuration, and necrosis of connective tissue, and often terminating in systemic infection.

Colpitis. Signs of vaginal inflammation, simple catarrhal, ulcerative, diphtheritic; labia often edematous in ulcerative vaginitis.

Pure septicemia. Characterized by pyrexia with absence of perceptible organic lesions; countenance sallow, sunken, anxious; occasionally delirium or coma; diarrhea, and vomiting of dark grumous ejecta; runs a rapid course, terminating within a few days.

In most cases of puerperal fever several of the lesions above described coexist.

Treatment.

Prophylactic. Prevent infection by careful disinfection of the hands, instruments, utensils, etc., before each contact with the genitals. Cleanse, antiseptically, the external genitals, lower abdomen, and inner surfaces of the thighs before examination. Disinfect the vagina and cervix before labor, for cause. Examine per vaginam, during labor, as seldom as possible. In most cases vaginal examination may, when special care is required, be omitted altogether. Avoid all preventable injury to the passages. The mortality from puerperal infection in private practice should be *nil.*

Remedial. General Treatment of Infection. Dislodge the enemy and reinforce the resisting powers of the patient.

On the first rise of temperature, give hydrarg. chlorid. mit., gr. x to gr. xx, and follow with a saline—Epsom salts.

Repeat the saline to procure at least three or four watery movements daily, if the strength of the patient permits. This treatment applies especially to the first few days of the fever.

Vaginal Douche. Douche the vagina with a 1 : 1000 hydronaphthol, or 1 : 10 chlorinated soda solution, or the peroxide of hydrogen in full or half strength. If the temperature falls after the douche, the irrigation should be repeated as soon as it begins to rise again.

Uterine Douche. If the pyrexia is not relieved within six hours by the vaginal disinfection, douche the uterine cavity with one of the non-mercurial antiseptic solutions, best the peroxide of hydrogen.

Curetting. This failing, the uterine cavity being septic, immediately remove all necrotic material with a large curette and antiseptic douche. Leave ten or twenty grains of iodoform in the uterus, and drain the cavity by means of a strip of iodoform gauze folded to an inch in width and pushed gently into the uterus. Remove the gauze drain in two or three days, sooner if it becomes in the least fetid.

Support the patient with tonics—iron, quinine, strychnine, stimulants (maximum dose nearly one quart of brandy or its equivalent daily), and forced alimentation. Reduce the temperature by cold sponging, cold packs, or the use of the cold coil.

Treatment of Peritonitis.—Hydragogue cathartics

with large stimulating enemata, to procure several copious evacuations daily, to be continued p. r. n. Moderate doses of opium will rarely be required, to control pain or restlessness. Dietetic supports, tonics, and stimulants are to be freely used. Remove necrotic material and septic fluids from the utero-vaginal tract by the douche or curette and douche. In localized purulent peritonitis, open the abdomen, irrigate, and drain the pus cavity.

Treatment of Parametritis.—Hot vaginal douches, several gallons, temp. 110° to 120° F., two or three times daily. Antiseptic and general treatment as above indicated. In pelvic abscess, evacute early and drain, by the vagina or the abdomen, as the indications in the case may require.

Treatment of Colpitis.—Irrigate several times daily with a two-and-a-half per cent. creolin mixture, a chlorinated-soda solution, one in ten, or with peroxide of hydrogen. ·Touch necrotic patches with liq. ferri perchlorid., tr. iodine, or with a fifty per cent. solution of zinc chlorid.

Treatment of Phlegmasia Alba Dolens.—Keep the limb at rest in a horizontal position. Subdue pain by the local application of morphinæ oleas. After the application envelope the limb with a single thickness of muslin wrung out of hot water, and cover this with oiled silk. Avoid massage during the active stage of the disease; it may cause embolism. The patient may leave the bed when the swelling subsides and the fever has long since ceased; from that time should use support by means of an elastic (flannel) bandage or elastic stocking.

The cellulitic form should be treated by early and free incisions of the diseased structures.

SUDDEN DEATH IN CHILDBED.

Among the causes of sudden death in childbed are shock, syncope, pulmonary embolism, air embolism, acute pulmonary edema, apoplexy, advanced cardiac lesions.

APPENDIX.

FORM FOR CASE RECORD.

Case of

No.   -   Date of application   18

I. History.

Name   Residence   Age   years
Nativity   Married   years   para
Character of previous pregnancies   labors
    puerperiums
Miscarriages
Last menses from   to   Quickened
Health during present pregnancy

II. Preliminary Examination
(A month before labor.)

General Condition.
Mammary Glands, development
    Nipples, healthy   well-developed or not
Abdominal Examination.
    Uterus, shape   direction   height of fundus
    Liquor amnii, excessive or not
    Location of placenta   "
    Complicating tumors
    Fetus, one, two
    Fetal dorsum to mother's front, back, right, left.
    Fetal head, location   size
    Fetal heart-tones, location   rate   rhythm
    Length of fetal ovoid

VAGINAL EXAMINATION.
    Pudendum
    Vagina
    Cervix, old injuries   size   consistence
PELVIC MEASUREMENTS.
    Inter-cristal   Inter-spinal
    External conjugate   Diagonal conjugate.
    True conjugate   Other measurements
URINARY EXAMINATION.
    Daily amount   Reaction
    Specific gravity   Albumin
    Sugar   Casts.
                          Other microscopic findings
SUBSEQUENT OBSERVATIONS.

III. LABOR.

First Stage.

PAINS began   frequency   character.
GENERAL CONDITION.
    pulse   temperature
ABDOMINAL EXAMINATION.—Items as in II.
      Also :
    Abdomen pendulous   Bladder full or empty
VAGINAL EXAMINATION.—Items as in II.
      Also :
    Rectum full or empty   vaginal secretions
    Os internum effaced
    Os externum, size   consistence of margin
        thickness of margin
    Membranes ruptured or not
    Bag of waters, shape   size
    Presentation and position
COMPLICATIONS.
TREATMENT.
DURATION.

Second Stage.

GENERAL CONDITION.
    pulse   temperature
CHARACTER OF PAINS.
VAGINAL SECRETIONS.
MEMBRANES ruptured when   how
CAPUT SUCCEDANEUM, size
MECHANISM.
PERINEAL STAGE, duration   management
COMPLICATIONS.
MANAGEMENT, medication
              operative interference
TERMINATED at
DURATION.

Third Stage.

UTERINE CONTRACTIONS.
PLACENTAL DELIVERY at   method
COMPLICATIONS.
PLACENTA.
    Length   Width
    Weight   Anomalies
UMBILICAL CORD, insertion
    length   anomalies.
MEMBRANES, complete or not   how removed
UTERUS, retraction   height of fundus   shape
INJURIES.
    Uterus   Cervix   Vagina
    Pudendum   Treatment.
GENERAL CONDITION.
    pulse   temperature
TREATMENT.
DURATION of placental stage.

IV. SUBSEQUENT DAILY RECORD.

GENERAL CONDITION.
    pulse   temperature
DIET.
BREASTS AND NIPPLES.

BOWELS.

BLADDER. (Examine over abdomen for overdistention.)

UTERUS, height of fundus   width

     consistence   sensitiveness

        (In third or fourth week examine bimanually.)

LOCHIA, amount   character

     color   odor

OTHER OBSERVATIONS.

TREATMENT.

Condition on Dismissal.

18

GENERAL CONDITION.

BREASTS.

UTERUS, size   shape   position

     Cervix, size   shape   position

     injuries   cervical canal, how large

VAGINA.

PUDENDUM.

OTHER PELVIC STRUCTURES.

V. CHILD.

Observations at Birth.

SEX.

GENERAL CONDITION.   respiration

     circulation   pulse

     temperature in rectum

     rectum and urethra pervious

DEVELOPMENT.

     length   weight

	O. M.	O. F.	S. O. B.	BI-P.	BI-T.
Head { diameters					
{ circumferences					

     Other measurements.

CAPUT SUCCEDANEUM, size   location

SKIN   vernix caseosa   lanugo

INJURIES.

CONGENITAL ANOMALIES.

Subsequent Daily Record.

GENERAL CONDITION.
EYES.
MOUTH.
SKIN.
DIGESTION.
UMBILICAL WOUND.
NUTRITION, breast   bottle
    well nourished or not
    weekly gain in weight
OTHER OBSERVATIONS.
TREATMENT.

INDEX.

ERRATA.

Page 20, line 7, *for* " layers," *read* " layer."

Page 21, lines 28 and 29, strike out " the *sphincter vaginæ.*"

Page 33, line 8, *for* " menstruation," *read* " the menses."

Page 101, line 27, *for* " fascia," *read* " fasciæ."

Page 135, line 1, *for* " posterior " shoulder, *read* " anterior " shoulder.

Page 164, line 20, *for* " general septicemia," *read* " septicemia."

Page 194, line 27, *for* " reflex," *read* " pelvic."

Page 201, line 4, strike out comma after " before," and insert after " ovariotomy."

www.ingramcontent.com/pod-product-compliance
Lightning Source LLC
Chambersburg PA
CBHW020850270326
41928CB00006B/633